Contents

CW01500009

Linear equations and matrices

Mathematics for Engineers

The series is designed to provide engineering students in colleges and universities with a mathematical toolkit, each book including the mathematics in an engineering context. Numerous worked examples, and problems with answers, are included.

1. Laplace and z-transforms
2. Ordinary differential equations
3. Complex numbers
4. Fourier series
5. Differentiation and integration
6. Linear equations and matrices

Mathematics for Engineers

Linear equations and matrices

W. Bolton

Longman
Scientific &
Technical

Longman Scientific & Technical
Longman Group Limited
Longman House, Burnt Mill, Harlow
Essex CM20 2JE, England
and Associated Companies throughout the world

First published 1995

British Library Cataloguing in Publication Data
A catalogue entry for this title is available from the British Library

ISBN 0–582–25633–X

Printed in Malaysia

Preface

This is one of the books in a series designed to provide engineering students in colleges and universities with a mathematical toolkit. In the United Kingdom it is aimed primarily at HNC/HND students and first-year undergraduates. Thus the mathematics assumed is that in BTEC National Certificates and Diplomas, the Advanced General National Vocational Qualification, or in A level. The pace of development of the mathematics has been aimed at the notional reader for whom mathematics is not their prime interest or 'best subject' but need the mathematics in their other studies. The mathematics is developed and applied in an engineering context with large numbers of worked examples and problems, all with answers being supplied.

This book is concerned with linear equations and matrices, with emphasis on the solution of simultaneous linear equations. A familiarity with basic algebra is assumed. The aim of the book has been to include sufficient worked examples and problems to enable the reader to acquire an understanding of techniques used and proficiency in the solution of simultaneous linear equations in engineering applications. Chapters on such equations in electric circuit analysis and structural analysis are included.

W. Bolton

1 Linear equations

1.1 Linear equations

This book is about linear equations and in particular the solution of engineering problems which involve simultaneous linear equations. In this chapter the definition of a linear equation is given and methods are considered that can be used to solve two equations involving two unknowns, three equations with three unknowns and, in general, n equations with n unknowns. In chapter 2 the concept of a matrix is considered. Matrices are a theoretical and practically useful way of approaching many types of problems, in particular the solution of simultaneous linear equations. Chapters 6 and 7 illustrate this use in electrical circuit analysis and the analysis of structures.

1.1.1 Defining a linear equation

The term *linear equation* is used for an equation which gives a straight line graph. A straight line graph has a constant slope. Thus, for the graph shown in figure 1.1(a) where the line passes through the points (x_1, y_1) and (x_2, y_2) the slope m is given by

$$m = \frac{y_2 - y_1}{x_2 - x_1}$$

If we take a linear graph which has an intercept with the y-axis of c, i.e. as shown in figure 1.1(b), then the line passes through the points $(0, c)$ and (x_2, y_2) and so the slope is given by

$$m = \frac{y_2 - c}{x_2 - 0}$$

This can be written as

$$y_2 = mx_2 + c$$

This is the *equation of the straight line* and is written in general as

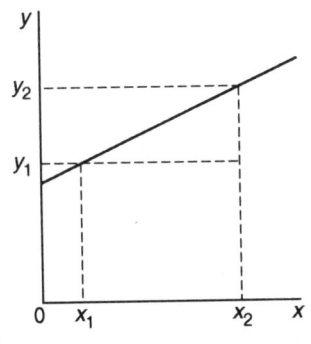

(a)

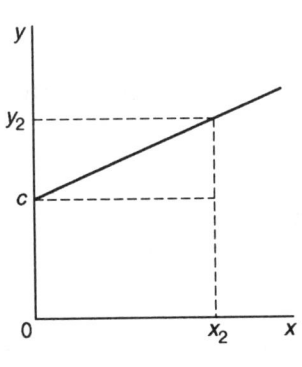

(b)

Fig. 1.1 Straight line graphs

$$y = mx + c \tag{1}$$

The equation of a straight line, i.e. a linear equation, is one in which the variables are only raised to the power +1 and does not involve any products of variables, e.g. xy, or the variables occurring in trigonometric, logarithmic or exponential functions. For example, a graph of $yx = 1$ does not describe a straight line graph. We can write this as $y = 1x^{-1}$ and so it contains a power other than +1. Likewise, a graph of $y = x^2$ does not describe a straight line graph but a curve and a graph of $y = \sin x$ does not describe a straight line but a sine wave.

Equation [1] describes a line in the xy-plane, such an equation being a linear equation in the variables x and y. There are only two variables. We can, however, have equations involving more than two variables. e.g. $y = ax + bz + c$. Such equations are linear if they follow the same conditions as outlined above. In general, a linear equation involving n variables is of the form

$$a_1 x_1 + a_2 x_2 + ... + a_n x_n = c \tag{2}$$

where a_1, a_2, ... a_n and c are constants and x_1, x_2, ... x_n are variables.

With all linear equations we can, without affecting the relationship between the variables, multiply both sides of the equation by the same constant. For example, $y = 2x + 1$ when multiplied by 2 gives $2y = 4x + 2$ and still gives the same relationship between y and x. It is still the same line on a graph.

Example

Which of the following equations are linear?

(a) $y = 3x + 2$, (b) $yx = 2$, (c) $y = 5$, (d) $y + 2z + 3x = 5$,

(e) $y = 2x^2 + 5$, (f) $x = 3y$, (g) $y = 2\,e^x$

(a) This equation is of the form given in equation [1], involving only variables to the power 1. It is thus linear.
(b) This equation involves the product of variables and is not linear.
(c) This equation is of the form given in equation [1] with $x = 0$. It is a linear equation describing a straight line parallel to the x-axis.
(d) This equation only involves variables to the power +1 and so is linear.
(e) This equation contains a variable to the power +2 and so is not linear.
(f) This equation is of the form given in equation [1] with $c = 0$. It is thus linear.

(g) This equation contains a variable as an exponential function and so is not linear.

Example

For one mesh of an electrical circuit, Kirchhoff's voltage law gives the equation $2I_1 + 5I_2 = 3$. Is the equation linear?

The equation involves variables only to the power +1 and is of the form given in equation [2]. The equation is thus linear. An electrical circuit which gives only linear equations is said to be linear.

Review problems

1 Which of the following are linear equations, x, y and z being variables?

(a) $y - \sin x = 0$, (b) $y + x = 7$, (c) $y = x^2 + 5$,

(d) $y = \frac{1}{2}x + 2z + 1$, (e) $x_1 + 2x_2 + x_3 = 5$, (f) $y = 2 + 3xz$

2 For a number of forces acting at a point the following relationships are obtained. Are the equations linear?

(a) $2F_1 + 3F_2 - 4F_3 = 12$, (b) $F_1 \cos 30° + F_2 \sin 60° = 10$

3 For an electrical circuit, Kirchhoff's voltage law applied to meshes in the circuit gave the following equations. Are the equations linear?

(a) $2 = 3I_1 + 2(I_1 - I_2)$, (b) $0 = 2(I_2 - I_1) + 5I_2$

1.2 Solutions

Consider a linear equation

$$x + 2y = 2$$

The term *solution* is used to describe values of x and y that satisfy the equation, i.e. values of x and y for points that lie on the straight line graph given by the equation. Thus when:

$x = 0$ the value of y that satisfies the equation is $y = 1$,

$x = 1$ the value of y that satisfies the equation is $y = 1/2$,

$x = 2$ the value of y that satisfies the equation is $y = 0$,

$x = 3$ the value of y that satisfies the equation is $y = -1/2$,

and so on.

We thus have, as solutions, all the values that lie on the straight line and so there is an infinite set of solutions.

Now consider the following system of two linear equations

$$x + 2y = 2$$

$$x - 3y = 7$$

If the two equations *simultaneously* apply then the values of x and y that satisfy the first equation must also satisfy the second equation. If we plot the straight lines for the two equations, then the point of intersection between the lines gives the values of x and y that satisfy both equations. Figure 1.2 shows the graphs. The solution for the simultaneous pair of equations is thus $x = 4$ and $y = -1$.

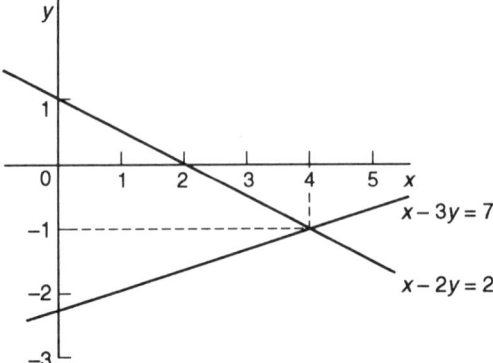

Fig. 1.2 The solution of $x + 2y = 2$ and $x - 3y = 7$

If we have two simultaneous linear equations that intersect, as above, then adding the equations also gives an equation that intersects at the same point. Thus $(x + 2y) + (x - 3y) = (2 + 7)$, i.e. $2x - y = 9$. Thus we can still have $x = 4$ and $y = -1$. This is an important property of linear equations. In engineering the property is used with electrical circuits and forces on structures in what is termed the *principle of superposition*.

Not all pairs of simultaneous equations have solutions, i.e. have lines that intersect on a graph. We could have a pair of lines that are parallel, for example

$$x + 2y = 2$$

$$x + 2y = 4$$

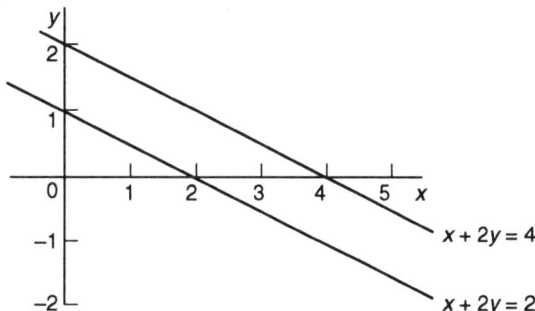

Fig. 1.3 The graphs of $x + 2y = 4$ and $x + 2y = 2$

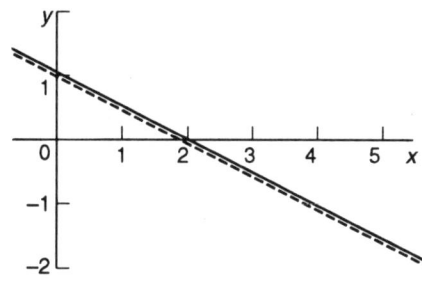

Fig. 1.4 The graphs of $x + 2y = 2$ and $2x + 4y = 4$

The graph of these lines is shown in figure 1.3.

Another possibility is that the pair of equations have the same graph. Then the lines coincide (figure 1.4) and we have an infinite number of points of intersection and so an infinite number of solutions. For example

$$x + 2y = 2$$

$$2x + 4y = 4$$

describe two lines which coincide.

Thus the possibilities with a pair of equations are:

1   The lines intersect at one point and so there is just one solution.
2   The lines are parallel and so there is no intersection and consequently no solution.
3   The lines coincide and so there are an infinite number of points of intersection and so an infinite number of solutions.

Review problems

4   By graphical means, determine whether the following sets of equations have one solution, no solution or an infinite number of solutions.

(a) $y = 2x + 1$, $y = 3x + 2$, (b) $2y + x = 3$, $2y + x = 6$,

(c) $y = 3x + 1$, $2y = 6x + 2$

1.3 Two equations with two unknowns

As indicated in the previous section, we can determine the solution of two equations with two unknowns, i.e. the values of the two unknowns that simultaneously fit both equations, by graphical means. In this section we consider how the equations can be solved by algebraic means. Chapter 2, on matrices, can be considered an extension of this.

1.3.1 Solution by elimination

Consider the two simultaneous linear equations

$$x + 2y = 2 \quad \text{and} \quad x - 3y = 7$$

There is a standard procedure which can be used to solve this pair of equations, or indeed any set of simultaneous equations, no matter how many equations are involved. The procedure involves taking a pair of equations, then *eliminating* one of the unknowns in the pair by combining the equations in a suitable manner. Thus, with the above pair of equations, if we want to eliminate x we can just subtract the two equations, i.e.

$$\begin{array}{r} x + 2y = 2 \\ \text{minus } \underline{x - 3y = 7} \\ 0 + 5y = -5 \end{array}$$

Thus $y = -1$. We can obtain the value of x by substituting this value of y into one of the equations. Alternatively, we can obtain x by the elimination of y between the equations. We can do this by multiplying the first equation by 3 and the second equation by 2, then adding.

$$\begin{array}{r} 3x + 6y = 6 \\ \text{plus } \underline{2x - 6y = 14} \\ 5x + 0 \;\; = 20 \end{array}$$

Thus $x = 4$.

Example

By using Kirchhoff's voltage law to determine the currents in a circuit consisting of two meshes the following equations were obtained:

$$6I_1 + 2I_2 = 12 \quad \text{and} \quad 4I_1 - 12I_2 = 0$$

with the currents in amperes. Determine the currents I_1 and I_2.

To eliminate I_2 between the two equations we multiply the first equation by 6 and then add the resulting equations.

$$36I_1 + 12I_2 = 72$$
$$\text{plus} \quad 4I_1 - 12I_2 = 0$$
$$40I_1 + 0 = 72$$

Thus $I_1 = 1.8$ A. Substituting this value in the first equation gives

$$6 \times 1.8 + 2I_2 = 12$$

Hence $I_2 = 0.6$ A.

Review problems

5 Determine, by means of elimination, the solutions of the following pairs of equations:

(a) $x - 3y = 4$, $- 4x + 2y = 6$, (b) $2y + 3x = 1$, $y - 2x = 4$,

(c) $y + 3x = -7$, $3y - 2x = 12$, (d) $x + 3y = 5$, $2x + y = 2$,

(e) $y - 2x = 4$, $2y + 3x = -6$, (f) $x + y = 5$, $6x - y = 9$

6 By using Kirchhoff's voltage law to determine the currents in a circuit consisting of two meshes the following equations were obtained:

$$4I_1 - 3I_2 = 1 \quad \text{and} \quad 3I_1 - 5I_2 = 1$$

with the currents in amperes. Determine the currents I_1 and I_2.

7 With a machine the effort E in newtons required to overcome a load L in newtons is given by the equation $E = a + bL$, where a and b are constants. It is found that an effort of 5 N is required to overcome a load of 8 N and an effort of 7 N to overcome a load of 12 N. Determine the values of the constants.

8 A calculation of the stresses in the steel bars and concrete of a reinforced concrete column leads to the following pair of equations:

$$200 \times 10^3 = 29\,700 \times 10^{-6}\sigma_c + 300 \times 10^{-6}\sigma_s$$

$$\frac{\sigma_c}{20 \times 10^9} = \frac{\sigma_s}{200 \times 10^9}$$

with the stresses σ_s and σ_c being in pascals. Determine the stresses.

1.3.2 Solution by substitution

A technique which is often used to eliminate an unknown from an equation of a pair is to rearrange one equation to isolate one of the unknown variables. This variable is then substituted into the other equation. Consider, for example, the pair of equations discussed earlier, namely

$$x + 2y = 2 \text{ and } x - 3y = 7$$

If we rearrange the first equation we can obtain $x = 2 - 2y$. If we now substitute the expression for x into the second equation we obtain

$$(2 - 2y) - 3y = 7$$

Hence $y = -1$. Substituting this value of y into either of the original equations gives $x = 4$.

Review problems

9   Solve, by substitution, the following pairs of equations:

   (a) $y + 2x = 10$, $2y - x = 5$, (b) $x + y = 2$, $2x - y = 7$,

   (c) $3x - y = 5$, $2x + y = 10$

1.4 *n* equations with *n* unknowns

In this section we consider the extension of the technique of elimination to the problem of solving n simultaneous equations involving n unknowns. The previous section was a consideration of this problem with $n = 2$. Chapter 2 takes the problem further and a technique called Gaussian elimination is developed, together with a form of notation called matrices.

   To solve three simultaneous equations we need to proceed in stages. First we take a pair of the three equations and eliminate one of the unknowns. We then eliminate the same unknown from a different pair of equations taken from the three. This results in two equations with two unknowns which can then be solved as detailed in the previous section. The procedure, for three equations 1, 2 and 3, can be thus be considered to follow the following stages:

1   Take equations 1 and 2 and eliminate one of the variables to produce an equation 4 in two variables.
2   Take equations 2 and 3, or 1 and 3, and eliminate the same variable as in stage 1 and produce an equation 5 in the same two variables as equation 4.

3   Take equations 4 and 5, a pair of equations with the same two variables, and eliminate one variable from them. The result is a value for one of the variables. Substitution of this value into either equation 4 or 5 can be used to obtain a second variable. Substitution of these two values into equation 1, 2 or 3 can then be used to give the third variable.

The following example illustrates this technique. The technique can be extended to n variables with n equations, but this is a matter which is taken up in more detail in chapter 2.

Example

Solve the following three simultaneous equations:

$$x + 2y + z = 2, \quad 2x + 3y - z = 5, \quad x - 3y + 2z = 1$$

Using the stages outlined above:

Stage 1
If we add the first and second equations we can eliminate z to obtain

$$\begin{array}{r} x + 2y + z = 2 \\ \text{plus } \underline{2x + 3y - z = 5} \\ 3x + 5y + 0 = 7 \end{array}$$

Stage 2
If we multiply the second equation by 2 and add it to the third equation then we eliminate z to obtain

$$\begin{array}{r} 4x + 6y - 2z = 10 \\ \text{plus } \underline{x - 3y + 2z = 1} \\ 5x + 3y + 0 = 11 \end{array}$$

Stage 3
We now have a pair of equations in x and y. If we multiply the first equation by 3, the second equation by 5, and then subtract one from the other, we obtain

$$\begin{array}{r} 25x + 15y = 55 \\ \text{minus } \underline{9x + 15y = 21} \\ 16x + 0 = 34 \end{array}$$

Thus $x = 2.125$. We can substitute this into one of the above equations in x and y to obtain $y = 0.125$. We can substitute these values of x and y in one of the original equations and so obtain $z = -0.375$.

Review problems

10 Solve the following three simultaneous equations:

$$2x + 4y + 6z = 18, \quad 3x + y - 2z = 4, \quad 4x + 5y + 6z = 24$$

11 Solve the following three simultaneous equations:

$$2x + 3y + z = 5, \quad x + 2y + 3z = 6, \quad 3x + y + 2z = 1$$

12 Kirchhoff's voltage law when applied to a circuit with three meshes gives the following three equations:

$$2I_1 + 8(I_1 - I_2) = 40, \quad 8(I_2 - I_1) + 6I_2 + 6(I_2 - I_3) = 0$$

$$6(I_3 - I_2) + 4I_3 = -20$$

with the three currents being in amperes. Determine the currents.

Further problems

13 Which of the following equations are linear?

(a) $x + y = 4$, (b) $xy = 4$, (c) $y = 1/x$, (d) $y = (x + 1)^2$,

(e) $y = 4 \cos x$, (f) $y = 2 \ln x$, (g) $x + 2y + 3z = 5$,

(h) $y = 2\sqrt{x} + 3$, (i) $2x_1 + x_2 - 3x_3 = 5$, (j) $4x - 2y = 5$

14 Which of the following pairs of equations have a single solution?

(a) $y = x + 2$, $y = x + 5$, (b) $y = x + 1$, $y = 3x + 1$,

(c) $y = 2x + 1$, $y - 2x = 1$, (d) $x + 2y = 3$, $2x - y = 1$

15 Determine, by means of elimination, the solutions of the following pairs of equations:

(a) $4y - x = 5$, $3y + 2x = 12$, (b) $2y + 3x = 12$, $3y + 4x = 5$,

(c) $y + 4x = 14$, $15y + 2x = 36$, (d) $y - x = 5$, $2y + x = 4$,

(e) $3y - 4x = 4$, $y + 3x = 10$, (f) $4y + 3x = 36$, $3y - 2x = 44$

16 In using Kirchhoff's voltage law to determine the currents in a

circuit consisting of two meshes the following equations were obtained:

$$4I_1 - I_2 = 26 \quad \text{and} \quad -I_1 + 3I_2 = -12$$

with the currents in amperes. Determine the currents I_1 and I_2.

17 As a result of resolving forces acting at a point the following equations are obtained:

$$T \cos 30° = F \cos 45° - 10 \cos 60° + 20 \cos 30°$$

$$T \sin 30° + 20 \sin 30° = F \sin 45° + 10 \sin 60°$$

with T and F being forces in newtons. Determine the values of the forces.

18 In using Kirchhoff's voltage law to determine the currents in a circuit consisting of two meshes the following equations were obtained:

$$5I_1 + 10(I_1 + I_2) = 8 \quad \text{and} \quad 10(I_1 + I_2) + 6I_2 = 10$$

with the currents in amperes. Determine the currents I_1 and I_2.

19 For an electrical circuit involving three meshes, Kirchhoff's voltage law gives, with the current in amperes,

$$14I_1 - 5I_2 - 3I_3 = 8, \quad 3I_1 + 4I_2 - 9I_3 = 0, \quad 5I_1 - 10I_2 + 4I_3 = -7$$

Determine the three currents.

2 Gaussian elimination

2.1 *n* equations with *n* unknowns

In the previous chapter three equations with three variables, say x, y and z, were solved by considering one pair of the equations, multiplying one of the equations throughout by a number so that when added to the other one of the pair it eliminates one of the variables. We might, for example, eliminate z between equations 1 and 2 and so obtain an equation with just x and y. This was then repeated for another pair of the three equations with the result that a pair of equations in just two variables was obtained. Thus, z might have been eliminated between equations 2 and 3 to give a second equation having just the variables x and y. The resulting pair of equations was then solved by elimination of one of the variables. Thus with three equations, we reduce the initial two pairs in three variables to one pair in two variables.

Now suppose we had, say, five simultaneous equations in five variables. We can consider the five equations to represent four pairs from which we can eliminate the same variable. The result is then four equations in four variables. These four can be considered to be three pairs from which we can eliminate the same variable. The result is then three equations in three variables. These three can be considered to be two pairs from which we can eliminate the same variable. The result is then two equations in two variables. This pair can then be solved by eliminating one of the variables.

As the number of equations and unknowns increases then the number of steps involved in the technique increases dramatically. To cope with this a routine process is required that can be systematically applied so that the process becomes essentially automatic, this particularly being required if simultaneous equations are to be solved by a computer. One such process is called *Gaussian elimination*. Karl Friedrich Gauss (1777–1855) was a notable mathematician and scientist.

2.1.1 Gaussian elimination

The Gaussian elimination procedure involves the following steps, being illustrated for three simultaneous equations [1], [2] and [3] involving the variables x_1, x_2, x_3:

1 Divide the first equation [1] by the coefficient of x_1. If the coefficient is zero then choose another of the equations and make it the first equation. The result is a new first equation [4].

2 Eliminate x_1 from the second equation [2] by subtracting a multiple of the new first equation [4] from the second equation. Eliminate x_1 from the third equation by subtracting a multiple of the new first equation [4] from them. The result is two new equations [5] and [6].

3 Divide equation [5] by the coefficient of x_2. If the coefficient is zero then choose equation [6] and reverse the sequence of the equations. This becomes equation [7].

4 Eliminate x_2 from equation [6] by subtracting a multiple of equation [7] from equation [6]. This becomes equation [8].

5 Divide equation [8] by the coefficient of x_3.

The result of the above elimination stages is that the value of x_3 can be read off from the last stage. This value can then be substituted into equation [5] to give x_2. These values of x_3 and x_2 are then substituted into equation [4] to give the value of x_1. The following example, involving the three simultaneous equations considered in the example in the previous chapter, illustrates the above process.

With two simultaneous equations and two variables the Gaussian elimination process requires three divisions, three multiplications and three subtractions. With three simultaneous equations and three variables it requires six divisions, eleven multiplications and eleven subtractions. With four simultaneous equations and four variables it requires ten divisions, twenty-six multiplications and twenty-six subtractions.

Example

Solve, using the Gaussian elimination process, the following three equations:

$$x + 2y + z = 2 \qquad\qquad [1]$$

$$2x + 3y - z = 5 \qquad\qquad [2]$$

$$x - 3y + 2z = 1 \qquad\qquad [3]$$

The following are the stages in the Gaussian elimination process:

Stage 1	$[1] \div 1$	$x + 2y + z = 2$	[4]
Stage 2	$[2] - 2 \times [4]$	$-y - 3z = 1$	[5]
	$[3] - 1 \times [4]$	$-5y + z = -1$	[6]
Stage 3	$[5] \div (-1)$	$y + 3z = -1$	[7]
Stage 4	$[6] - (-5) \times [7]$	$16z = -6$	[8]
Stage 5	$[8] \div 16$	$z = -3/8$	

Thus $z = -3/8$. Substituting this value into equation [5] gives $y = 1/8$. Substituting these values into equation [4] gives $x = 17/8$.

Review problems

1   Use Gaussian elimination to solve the following pair of simultaneous equations:

$$x + 3y = 2, \quad 2x - y = 11$$

2   Use Gaussian elimination to solve the following three simultaneous equations:

$$2x - y + z = 6, \quad 4x + y + 3z = 10, \quad 3x - 5y - 2z = 9$$

2.2 A rectangular array

Consider the state of the three equations at the various stages in the Gaussian elimination process applied to the above example.

Initially we have

$$x + 2y + z = 2 \tag{1}$$

$$2x + 3y - z = 5 \tag{2}$$

$$x - 3y + z = 1 \tag{3}$$

Stage 1 gives

$$x + 2y + z = 2 \tag{4}$$

$$2x + 3y - z = 5 \tag{2}$$

$$x - 3y + 2z = 1 \tag{3}$$

Stage 2 gives

$$x + 2y + z = 2 \qquad\qquad [4]$$

$$-y - 3z = -1 \qquad\qquad [5]$$

$$-5y + z = -1 \qquad\qquad [6]$$

Stage 3 gives

$$x + 2y + z = 2 \qquad\qquad [4]$$

$$y + 3z = 1 \qquad\qquad [7]$$

$$-5y + z = -1 \qquad\qquad [6]$$

Stage 4 gives

$$x + 2y + z = 2 \qquad\qquad [4]$$

$$y + 3z = 1 \qquad\qquad [7]$$

$$16z = -6 \qquad\qquad [8]$$

Stage 5 gives

$$x + 2y + z = 2 \qquad\qquad [4]$$

$$y + 3z = 1 \qquad\qquad [7]$$

$$z = -3/8$$

After stage 1 the first of the three equations remains unchanged. After stage 3 the second equation remains unchanged. We are proceeding down the equations, fixing each one in turn into a new form. This applies regardless of how many equations are involved. The final form, stage 5 above, can be written as

$$x + 2y + z = 2 \qquad\qquad [4]$$

$$y + 3z = 1 \qquad\qquad [7]$$

$$z = -3/8$$

The equations have been reduced to a triangular format. This type of triangular form always occurs with Gaussian elimination.

Since all we are interested in at each stage are the coefficients of the variables, there is no need to write anything other than the coefficients. Thus, writing the coefficients in an array so that the first column indicates the coefficient of x_1, the second that of x_2, etc. and the first row is the coefficients of the first equation, the second row the second equation, etc., then each of the stages in the above example looks like:

Initially	1	2	1	2
	2	3	−1	5
	1	−3	2	1

Stage 1	1	2	1	2
	2	3	−1	5
	1	−3	2	1

Stage 2	1	2	1	2
	0	−1	3	1
	0	−5	1	−1

Stage 3	1	2	1	2
	0	1	3	1
	0	−5	1	−1

Stage 4	1	2	1	2
	0	1	3	1
	0	0	16	−6

Stage 5	1	2	1	2
	0	1	3	1
	0	0	1	$-\frac{3}{8}$

The operations on equations have now become operations on rows in an array and the stages can be stated as:

1 Divide the first row by the coefficient of x_1.
2 Subtract a multiple of the new first row from the second row to give a 0 in the first column. Subtract a multiple of the new first row from the third row to give a 0 in the first column.
3 Divide the new second row by the coefficient of x_2.
4 Subtract a multiple of the new second row from the third row to give a zero in the second column.
5 Divide the new third row by the coefficient of x_3.

The final array should have a diagonal of 1s down from the left top corner with a triangle of 0s in the lower left corner. This is just the triangular format of the equations referred to earlier in this section.

2.2.1 Matrices

The rectangular array of numbers we have used above to represent the coefficients of the simultaneous equations is called a *matrix*, plural *matrices*. The *coefficient matrix* is the array produced by just using the coefficients and the *augmented matrix* is one which has both the coefficients and the constant terms. Thus for the equations

$$x + 2y + z = 2 \qquad\qquad [1]$$

$$2x + 3y - z = 5 \qquad\qquad [2]$$

$$x - 3y + 2z = 1 \qquad\qquad [3]$$

we can write the coefficient matrix as

$$\begin{bmatrix} 1 & 2 & 1 \\ 2 & 3 & -1 \\ 1 & -3 & 2 \end{bmatrix}$$

and the augmented matrix is

$$\begin{bmatrix} 1 & 2 & 1 & 2 \\ 2 & 3 & -1 & 5 \\ 1 & -3 & 2 & 1 \end{bmatrix}$$

Note that brackets [ ] have been used to contain the data which form the matrix. The Gaussian elimination process is designed to end up with a matrix which has rows in which the first non-zero coefficient is a 1 and this 1 in any row is to the right of the first non-zero coefficient in the preceding row. Thus, in the example, the augmented matrix arrived at is

$$\begin{bmatrix} 1 & 2 & 1 & 2 \\ 0 & 1 & 3 & 1 \\ 0 & 0 & 1 & -\frac{3}{8} \end{bmatrix}$$

Such a form of matrix is called *row-echelon form*.

In carrying out Gaussian elimination, problems can arise. If the coefficient of a variable that is to be selected to be divided at a particular stage is zero then the sequence of the equations needs to be changed in order that the division can occur. When working the process through by hand we can choose which equation to interchange. When the elimination process is handled by a computer the rule that is adopted is that the interchange occurs with the row having the coefficient which is numerically the largest, negative signs being disregarded. This is known as *partial pivoting*. This has an advantage that the answers obtained for the variables can be made more accurately (see section 2.4.1). Problems also occur when there are no solutions or an infinite number of solutions (see section 2.5).

Example

Solve the following set of three linear equations by Gaussian elimination:

$$2x + y - z = 2, \quad x + 2y + z = 4, \quad 3x - y + 3z = 1$$

The augmented matrix is

$$\begin{bmatrix} 2 & 1 & -1 & 2 \\ 1 & 2 & 1 & 4 \\ 3 & -1 & 3 & 1 \end{bmatrix}$$

For stage 1 we make the first row start with a 1. This involves dividing that row by 2.

$$\begin{bmatrix} 1 & \frac{1}{2} & -\frac{1}{2} & 1 \\ 1 & 2 & 1 & 4 \\ 3 & -1 & 3 & 1 \end{bmatrix}$$

For stage 2 we make the second row start with a 0 by subtracting a multiple of the first row from it, in this case the multiple being $\times 1$. We also make the third row start with a zero by subtracting a multiple of the first row from it, in this case the multiple being $\times 3$. The result is then

$$\begin{bmatrix} 1 & \frac{1}{2} & -\frac{1}{2} & 1 \\ 0 & \frac{3}{2} & \frac{3}{2} & 3 \\ 0 & -\frac{5}{2} & \frac{9}{2} & -2 \end{bmatrix}$$

For stage 3 we make the second row have a 1 in its second column by multiplying the row by 2/3. The resulting matrix is then

$$\begin{bmatrix} 1 & \frac{1}{2} & -\frac{1}{2} & 1 \\ 0 & 1 & 1 & 2 \\ 0 & -\frac{5}{2} & \frac{9}{2} & -2 \end{bmatrix}$$

For stage 4 we make the third row have a 0 in the second column by subtracting from it a multiple of the second row, in this case the multiple being −5/2. The resulting matrix is

$$\begin{bmatrix} 1 & \frac{1}{2} & -\frac{1}{2} & 1 \\ 0 & 1 & 1 & 2 \\ 0 & 0 & 7 & 3 \end{bmatrix}$$

For stage 5 we make the third row have a 1 in the third column by multiplying the row by a number, in this case 1/7. The resulting matrix is

$$\begin{bmatrix} 1 & \frac{1}{2} & -\frac{1}{2} & 1 \\ 0 & 1 & 1 & 2 \\ 0 & 0 & 1 & \frac{3}{7} \end{bmatrix}$$

The final matrix thus reveals that $z = 3/7$. To obtain y we can substitute this value into the equation responsible for the second row. However, we can obtain this value by operations on the matrix. What we want is a row of the form

0   1   0   ?

This would describe an equation of the form $y = ?$ To obtain this form of row we can subtract the third row from the second row. The resulting matrix is then

$$\begin{bmatrix} 1 & \frac{1}{2} & -\frac{1}{2} & 1 \\ 0 & 1 & 0 & \frac{11}{7} \\ 0 & 0 & 1 & \frac{3}{7} \end{bmatrix}$$

Thus $y = 11/7$. We can obtain x by getting the first row into the form

1   0   0   ?

this describing an equation of the form $x = ?$ To obtain this form of row we can subtract half the second row and add half the third row. The resulting matrix is

$$\begin{bmatrix} 1 & 0 & 0 & \frac{3}{7} \\ 0 & 1 & 0 & \frac{11}{7} \\ 0 & 0 & 1 & \frac{3}{7} \end{bmatrix}$$

Thus $x = 3/7$.

The results can be checked by back-substituting the values into the original equations.

Example

Use the Gaussian elimination process to solve the following set of three equations:

$$x + 2y + z = 5, \quad 2x + 4y + z = 11, \quad x - y + 2z = 1$$

The initial matrix is

$$\begin{bmatrix} 1 & 2 & 1 & 5 \\ 2 & 4 & 1 & 11 \\ 1 & -1 & 2 & 1 \end{bmatrix}$$

The coefficient of x is already 1 so that the matrix is in the stage 1 format. For stage 2, when the first row is multiplied by 2 and subtracted from the second row, and when the first row is multiplied by 1 and subtracted from the third row, the matrix is as follows:

$$\begin{bmatrix} 1 & 2 & 1 & 5 \\ 0 & 0 & -1 & 1 \\ 0 & -3 & 1 & -4 \end{bmatrix}$$

For stage 3 we need to interchange the second and third rows since the second column of the row is a 0. Thus the matrix is now

$$\begin{bmatrix} 1 & 2 & 1 & 5 \\ 0 & -3 & 1 & -4 \\ 0 & 0 & -1 & 1 \end{bmatrix}$$

Dividing the second row by -3 gives

$$\begin{bmatrix} 1 & 2 & 1 & 5 \\ 0 & 1 & -\frac{1}{3} & \frac{4}{3} \\ 0 & 0 & -1 & 1 \end{bmatrix}$$

The third row is already in the required form for stage 4 so we proceed to stage 5. For stage 5 we divide the third row by -1.

$$\begin{bmatrix} 1 & 2 & 1 & 5 \\ 0 & 1 & -\frac{1}{3} & \frac{4}{3} \\ 0 & 0 & 1 & -1 \end{bmatrix}$$

Hence $z = -1$. If we add a third of row three to row two we obtain

$$\begin{bmatrix} 1 & 2 & 1 & 5 \\ 0 & 1 & 0 & 1 \\ 0 & 0 & 1 & -1 \end{bmatrix}$$

Thus $y = 1$. If we subtract twice the second row and also subtract the third row we obtain

$$\begin{bmatrix} 1 & 0 & 0 & 4 \\ 0 & 1 & 0 & 1 \\ 0 & 0 & 1 & -1 \end{bmatrix}$$

Hence $x = 4$.

The results can be checked by back-substituting the values into the original equations.

Review problems

3   Write the augmented matrix for the following sets of three linear equations:

(a) $x + 3y + 5z = 2$, $\ 2x - y + z = 4$, $\ 3x + 2y - z = 3$,

(b) $x + y + 2z = 1$, $\ -x + 3y + z = 2$, $\ 2x + y - z = 5$

4   Write the augmented matrix for the following set of four linear equations:

$2a + b + 3c - d = 2$, $\ -a + 3b + c + 2d = 1$,

$3a - b + c + 3d = 4$, $\ a + 2b - 3c - d = 1$

5   Write the linear equations represented by the following augmented matrix for the variables x, y and z:

$$\begin{bmatrix} 2 & 1 & 3 & 6 \\ -1 & 0 & 5 & 2 \\ 4 & -2 & -1 & 1 \end{bmatrix}$$

6   Use the Gaussian elimination process to solve the following sets of three equations:

(a) $x + 2y + 2z = 3$, $\ 2x - y + 3z = -2$, $\ -4x - 2y + z = 21$,

(b) $x + 3y - 2z = 11$, $\ 2x + 3y - z = 10$, $\ 4x - y + z = 4$,

(c) $x + y + 2z = 7$, $2x - 3y + 4z = 4$, $-x + y + z = 0$,

(d) $y + 3z = 5$, $2x - y - z = 9$, $x + 2y + z = 5$

7 Use the Gaussian elimination process to solve the following set of four equations:

$$3b + c - d = 4, \quad a + b - c + 2d = 0,$$

$$2a - b + c + d = 0, \quad a + 2b + 2c - d = 4$$

2.3 Residuals

If, after determining the solutions to a set of equations, the values are substituted back into each of the equations then ideally we should end up with $0 = 0$ for each equation. However this might not be the case; there may be some *residual*, i.e. difference between the values of the sum of the variables as expressed in an equation and the constant. This can occur if the values used in the rows and solutions are not exact, e.g. a calculator is used to obtain the values in the rows to say three decimal places and this has meant some rounding up or down of the values in order to fit the three decimal places. If this rounding up or down occurs in the calculation of values in an early row then, because these values are used in the calculation of later rows there will be an increase in error for those rows. The following example illustrates this.

Consider the following set of simultaneous equations and the determination of the solution when the calculations at each stage are carried out to three decimal places.

$$3x - y + 2z = 12, \quad x + 2y + 3z = 9, \quad 2x - 2y - z = 1$$

The augmented matrix is thus

$$\begin{bmatrix} 3 & -1 & 2 & 12 \\ 1 & 2 & 3 & 9 \\ 2 & -2 & -1 & 1 \end{bmatrix}$$

Dividing the first row by 3 gives

$$\begin{bmatrix} 1 & -0.333 & 0.667 & 4 \\ 1 & 2 & 3 & 9 \\ 2 & -2 & -1 & 1 \end{bmatrix}$$

Subtracting the first row from the second row and twice the first row from the third row gives

$$\begin{bmatrix} 1 & -0.333 & 0.667 & 4 \\ 0 & 2.333 & 2.333 & 5 \\ 0 & -1.334 & -2.334 & -7 \end{bmatrix}$$

Dividing the second row by 2.333 gives

$$\begin{bmatrix} 1 & -0.333 & 0.667 & 4 \\ 0 & 1 & 1 & 2.143 \\ 0 & -1.334 & -2.334 & -7 \end{bmatrix}$$

Subtracting −1.334 times the second row from the third row gives

$$\begin{bmatrix} 1 & -0.333 & 0.667 & 4 \\ 0 & 1 & 1 & 2.143 \\ 0 & 0 & 3.668 & 9.859 \end{bmatrix}$$

Dividing the third row by 3.668 gives

$$\begin{bmatrix} 1 & -0.333 & 0.667 & 4 \\ 0 & 1 & 1 & 2.143 \\ 0 & 0 & 1 & 2.688 \end{bmatrix}$$

Thus $z = 2.688$. Subtracting the third row from the second row gives

$$\begin{bmatrix} 1 & -0.333 & 0.667 & 4 \\ 0 & 1 & 0 & -0.545 \\ 0 & 0 & 1 & 2.688 \end{bmatrix}$$

Thus $y = -0.545$. If we multiply the second row by 0.333 and the third row by −0.667 and add these both to the first row we obtain

$$\begin{bmatrix} 1 & 0 & 0 & 2.026 \\ 0 & 1 & 0 & -0.545 \\ 0 & 0 & 1 & 2.688 \end{bmatrix}$$

Thus $x = 2.026$.

   If we back-substitute these values into the original equations we obtain the following residuals: for $3x - y + 2z = 12$, for which we should have $3x - y + 2z - 12 = 0$,

$$6.078 + 0.545 + 5.376 - 12 = 0.001$$

for $x + 2y + 3z = 9$, for which we should have $x + 2y + 3z - 9 = 0$,

$$2.026 - 1.090 + 8.064 - 9 = 0$$

for $2x - 2y - z = 1$, for which we should have $2x - 2y - z - 1 = 0$,

$$4.052 + 1.090 - 2.688 - 1 = 1.454$$

The residuals are thus 0.001, 0 and 1.454.

Review problems

8 Determine the solutions and consequential residuals for the following equations when Gaussian elimination is carried out to two decimal places:

$$1.20x + 2.34y = 5.24, \quad 2.43x - 1.20y = 3.25$$

9 Determine the solutions and consequential residuals for the following equations when Gaussian elimination is carried out to two decimal places:

$$1.00x + 0.99y = 2.00, \quad 0.99x + 1.00y = 2.00$$

2.3.1 Ill-conditioned equations

With some sets of equations, their solution is extremely sensitive to small changes in the coefficients. Such changes may be as a result of rounding up or down in calculation or just experimental error in the specification of the coefficient. For example, in electrical network analysis using Kirchhoff's laws the coefficients can be determined by the values of resistances. These are likely to have some tolerance associated with their values and so there is a margin of error associated with the coefficients. One way we can visualise this is to consider a system of two equations with two variables. The two equations can be represented by straight line graphs, the solution being their point of intersection (see section 1.2). For equations which are not sensitive to small changes in the coefficients we have graphs of the form shown in figure 2.1(a). A slight change in the slopes of the graphs will have little effect on

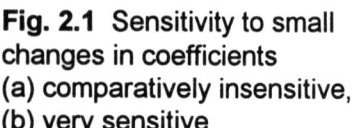

Fig. 2.1 Sensitivity to small changes in coefficients (a) comparatively insensitive, (b) very sensitive

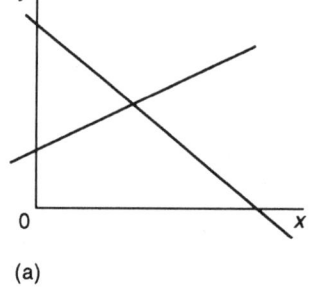

(a)

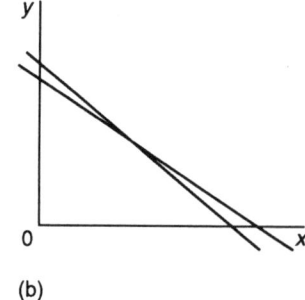

(b)

the point of intersection. For equations which are very sensitive to small changes in the coefficients then the lines are almost parallel, as in figure 2.1(b). A small change in a coefficient, such as a slight change in the slope of a graph, can then change the point of intersection considerably and hence the solution.

Equations for which small changes in the coefficients, as a result of inexact data or rounding up or down in calculation, result in large changes in solutions are said to be *ill-conditioned*. A pair of ill-conditioned equations are ones for which the two lines are nearly parallel.

Review problems

10 Consider the following set of equations:

$$x + y = 1, \ x + cy = 0$$

What will be the effect on the solutions if the coefficient c is determined as being between 1.1 and 1.2?

2.4 Row operations

The operations that can be carried out on a row in a matrix are:

1 Multiply, or divide, all the terms in a row by a non-zero constant.
2 Subtract, or add, a multiple of one row from/to another.
3 Interchange rows.

The first of these operations is just stating that the equality of an equation remains unchanged if it is multiplied throughout by the same non-zero constant. The second operation is merely stating that subtracting equal quantities from, or adding equal quantities to, both sides of an equation does not change the equality of the equation. Interchanging rows is merely changing the sequence in which equations are considered. Since the sequence was arbitrary in the first instance there is obviously no effect. These operations thus do not change the relationships represented by a set of equations and hence an augmented matrix.

2.4.1 Pivoting

The term *pivot* is used for the coefficient used in multiplying or dividing a row in the Gaussian elimination process. The pivot coefficient is the first coefficient following a zero in a row, or in the case of the first equation the first coefficient. If large numbers of equations have to be solved, the errors introduced by rounding up or down may well become very large as a result of a build-up of errors at each stage of row manipulation. Such errors can be reduced by selecting for a row operation the equation with the pivot having the largest magnitude, ignoring any negative signs. We are then always dividing by the largest number. This reduces the error. This process is known as *partial pivoting*.

For example, we have the augmented matrix

$$\begin{bmatrix} 1 & -1 & -5 & 7 \\ 3 & 4 & 1 & 9 \\ 2 & 1 & 3 & 0 \end{bmatrix}$$

For the first stage of the Gaussian elimination we are going to divide the first row by the coefficient in the first column. The row with the largest coefficient in the first column, i.e. the largest pivot, is the second row. We thus interchange the first and second rows, i.e. make the second row the pivot row, and so the matrix becomes

$$\begin{bmatrix} 3 & 4 & 1 & 9 \\ 1 & -1 & -5 & 7 \\ 2 & 1 & 3 & 0 \end{bmatrix}$$

Then, carrying out the first stage of Gaussian elimination, dividing the new first row by 3 we have

$$\begin{bmatrix} 1 & 1.33 & 0.33 & 3 \\ 1 & -1 & -5 & 7 \\ 2 & 1 & 3 & 0 \end{bmatrix}$$

For the next stage we subtract a suitable multiple of the first row from the second and third rows. The resulting matrix is then

$$\begin{bmatrix} 1 & 1.33 & 0.33 & 3 \\ 0 & -2.33 & -5.33 & 4 \\ 0 & -1.66 & 2.34 & -6 \end{bmatrix}$$

The next stage involves the coefficient in the second column. The second row has the largest value of that coefficient and so there is no need to interchange the second and third rows. Thus we have

$$\begin{bmatrix} 1 & 1.33 & 0.33 & 3 \\ 0 & 1 & 2.29 & -1.72 \\ 0 & -1.66 & 2.34 & -6 \end{bmatrix}$$

The process then continues with the subtraction from the third row of a suitable multiple of the second row.

At each stage in the elimination process, when a particular coefficient has to be selected for use in row multiplication or division, the row with the largest coefficient is chosen.

Review problems

11 Use partial pivoting to obtain the solution to the following set of simultaneous equations, rounding up or down to two decimal places:

$$0.03x + 0.02y + 1.00z = 1.05, \quad 0.50x + 1.00y - 0.50z = 1.00,$$

$$-0.01x + 0.03y + 1.00z = 1.02$$

12 Use partial pivoting to obtain the solution to the following set of simultaneous equations:

$$2x - 5y + 4z = 5, \quad 3x - 2y + 3z = 6, \quad x - y + z = 1$$

2.5 Existence of solutions

There are three possible outcomes when solving simultaneous linear equations (see section 1.2 for this discussion in relation to the graphs of two equations involving two variables):

1 There is a unique solution.
2 There is no solution.
3 There are an infinite number of solutions.

A set of simultaneous equations that has no solution is said to be *inconsistent*. If there is at least one solution then it is said to be *consistent*. This means that if the values for the variables are substituted in one equation they will provide the solution and consistently will do the same for each of the other equations.

In carrying out Gaussian elimination involving, say, three simultaneous equations and three variables x, y, z then if the last row of the augmented matrix after all the stages reads

0 0 1 c

then $z = c$ and there is thus either a unique solution or an infinite number of solutions to the system of equations. If the last row reads

0 0 0 c

then there is no solution since $c \neq 0$. This indicates that two, or more, of the simultaneous equations are incompatible with each other. For example, we might have the contradictory equations of $x + y + z = 1$ and $2x + 2y + 2z = 3$.

If there is a row in the augmented matrix which is entirely zeros, i.e.

0 0 0 0

then this row is of no help in determining the solution. Such a row occurs when a row is just a multiple of the row subtracted from it. This means that we have in essence only one equation relating the variables, rather than two. For example, we might have

$$x + 2y + z = 1 \quad \text{and} \quad 2x + 4y + 2z = 2$$

These will give the rows

$$\begin{bmatrix} 1 & 2 & 1 & 1 \\ 2 & 4 & 2 & 2 \end{bmatrix}$$

Thus if we subtract twice the first row from the second row we have

$$\begin{bmatrix} 1 & 2 & 1 & 1 \\ 0 & 0 & 0 & 0 \end{bmatrix}$$

When this occurs, it does not necessarily mean there is no solution or an infinite number of solutions, merely that we have insufficient equations, only two, to determine the three variables.

Example

Determine whether there will be a unique solution, no solution or an infinite number of solutions to the following set of simultaneous equations:

$$3x + 6y = 9, \quad 2x + 4y = 6$$

The second equation is just two-thirds of the first equation. So, since it is a multiple of the first equation, we effectively have only one equation relating the variables. There are thus an infinite number of solutions. The augmented matrix for the equations is

$$\begin{bmatrix} 3 & 6 & 9 \\ 2 & 4 & 6 \end{bmatrix}$$

Thus if we carry out a Gaussian elimination process, multiplying the first row by 1/3 gives

$$\begin{bmatrix} 1 & 2 & 3 \\ 2 & 4 & 6 \end{bmatrix}$$

and then subtracting twice the first row from the second gives

$$\begin{bmatrix} 1 & 2 & 3 \\ 0 & 0 & 0 \end{bmatrix}$$

The all zero row clearly indicates that there are an infinite number of solutions.

Review problems

13 Determine whether there will be a unique solution, no solution or an infinite number of solutions to the following sets of simultaneous equations:

(a) $2x + y = 3$, $x + 2y = 3$, (b) $3x - 9y = 12$, $-2x + 6y = -8$,

(c) $x - y = 1$, $2x - 2y = 1$

2.5.1 *m* equations with *n* unknowns

Consider a situation where we have more unknowns than there are equations, e.g. three unknowns but only two simultaneous equations. There are then insufficient equations to enable unique values of the unknowns to be obtained. What we can do in such a situation is let one of the unknowns take some value, say, t. We can then use the equations to determine the values of the other unknowns in terms of t. The following example illustrates this. Since t can be any value then there must be an infinite number of solutions.

If we have more equations than unknowns, e.g. two unknowns and three equations, then we have more equations than are necessary to determine the values of the unknowns. We might thus use the first and second of the equations to determine the values and try them in the third equation to ensure that the values fit all three equations. If the solution is *consistent* then the solution for the first two equations will also be the solution for the extra equation.

Simultaneous equations can only have a unique solution if there are at least as many equations as there are unknowns.

Example

Determine the solutions of the following set of simultaneous equations:

$$a + 3b - 5c = 4, \quad 2a + 5b - 2c = 6$$

There are three variables and two equations. Thus there will be an infinite number of solutions. The augmented matrix for the equations is

$$\begin{bmatrix} 1 & 3 & -5 & 4 \\ 2 & 5 & -2 & 6 \end{bmatrix}$$

When twice the first row is subtracted from the second row we have

$$\begin{bmatrix} 1 & 3 & -5 & 4 \\ 0 & -1 & 8 & -2 \end{bmatrix}$$

Let $z = t$, then the second row gives

$$-y + 8t = -2$$

and so $y = 2 + 8t$. The first row gives

$$x + 3(2 + 8t) - 5t = 4$$

Thus $x = -2 - 19t$.

Review problems

14 Determine the solutions, if possible, of the following set of simultaneous equations:

(a) $2x - y - z = 6, \quad 3x + y - 4z = -1,$

(b) $x + 2y = 5, \quad 2x + 3y = 7, \quad 3x - y = 4,$

(c) $x + y + z = 6, \quad x + 2y - z = 2$

Further problems

15 Use the Gaussian elimination process to solve the following sets of equations:

(a) $x + y + z = 4, \quad 4x + 2y - 3z = 33, \quad 2x - 3y + 2z = -2,$

(b) $x + 3y + 6z = 2, \quad 5x + 3y + 4z = 15, \quad 4x + 8y + 12z = 16,$

(c) $2x + y + z = 8$, $3x - 2y - 4z = -4$, $5x - 3y - 6z = -5$,

(d) $a + 2b - c + 3d = 6$, $2a - b + c + d = 7$,

   $a + b + c - 2d = -2$, $3a - 2b + 2c + d = 9$,

(e) $3y - z = 4$, $x + y + z = 2$, $2x - y + 2z = 1$,

(f) $3x - 3y + z = 1$, $2x + y - 3z = 0$, $-x + y + 2z = 2$

16 The forces F_1, F_2 and F_3, in newtons, in a framework in equilibrium are related by the following equations:

$$3F_1 - 2F_2 \cos 60° + 2F_3 = 12$$

$$4F_1 \sin 30° - 2F_2 - F_3 = -2 \cos 60°$$

$$2F_1 \cos 60° + 2F_2 + 6F_3 \sin 30° = 12$$

Determine the values of the forces.

17 The currents I_1, I_2 and I_3 occurring in a circuit having three meshes are given by

$$5I_1 - I_2 - I_3 = 8, \quad -I_1 + 4I_2 - I_3 = 1, \quad -I_1 - I_2 + 8I_2 = 5$$

The currents are in amperes. Determine their values.

18 The currents i_1, i_2 and i_3 in the star-connection shown in figure 2.2 are given by the simultaneous equations

$$Z_1 i_1 - Z_2 i_2 = e_1 - e_2, \quad Z_2 i_2 - Z_3 i_3 = e_2 - e_3, \quad i_1 + i_2 + i_3 = 0$$

Solve the equations for the three currents.

19 Use the Gaussian elimination process to solve, if possible, the following sets of equations:

(a) $x + 2y + z = 4$, $2x + 3y + z = 6$,

(b) $x - y + 3z = 2$, $x + 2y + z = 6$, $2x - y - z = -1$,

   $x + 3y + 2z = 7$,

(c) $x + y - 2z = 1$, $2x + 3y - 5z = 1$

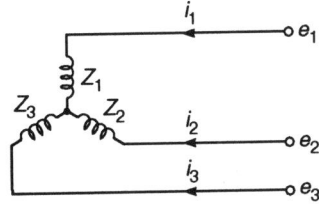

Fig. 2.2 Problem 18

3 Matrix operations

3.1 Matrices

A *matrix* is just a rectangular array of numbers. These can be real or complex numbers. In chapter 2 matrices were considered as a shorthand way of listing the coefficients of systems of linear equations and so make it easy to use the Gaussian elimination method to solve the equations.

An example of a matrix is that produced by the application of Kirchhoff's voltage law to an electrical circuit, a linear equation being produced for each mesh of that circuit. Thus we might have

$$5I_1 - I_2 - I_3 = 8$$

$$-I_1 + 4I_2 - I_3 = 1$$

$$-I_1 - I_2 + 8I_3 = 5$$

We can represent the coefficients of the currents as a rectangular array of numbers, i.e. a matrix

$$\begin{bmatrix} 5 & -1 & -1 \\ -1 & 4 & -1 \\ -1 & -1 & 8 \end{bmatrix}$$

If we add a column for the constants we have the *augmented matrix*, as used in chapter 2.

Another example can occur in a consideration of the stress distribution in a solid. A useful way of considering a solid is as being represented by a large number of interlinked elastic springs. We can then write equations for the stretching of each spring. The result is a large number of linear equations which can be represented by a matrix (see chapter 7). In control engineering, matrices are frequently used to represent sets of linear equations. Many engineering problems can be formulated in terms of a set of simultaneous equations, with the number in some instances being very large. Stating the equations in a matrix form gives a compact representation of the equations and can simplify the manipulation.

Though this book is primarily concerned with matrices and linear equations, matrices have uses that extend beyond this. There are many instances where data are arranged as a regular array of numbers, e.g. in any two-way table of data. For example, a spreadsheet indicating family expenditure is such an array, the following being a possible segment of such an array:

	January	February	March
Food	150	140	160
Telephone	20	18	22
Car and petrol	90	160	80
Clothes	20	30	100

In matrix notation this would be

$$\begin{bmatrix} 150 & 140 & 160 \\ 20 & 18 & 22 \\ 90 & 160 & 80 \\ 20 & 30 & 100 \end{bmatrix}$$

This chapter is about the terminology associated with matrices and the rules for the operations, such as addition and multiplication, that can be carried out with them. The aim is to deal with the basic principles of matrices and their use in relation to sets of simultaneous equations. The number of equations in such a set is kept relatively small for ease of handling 'by hand'; the principles are, however, applicable to the very large number of simultaneous equations that might be needed to represent some modern engineering applications, though computers would then be used to carry out the operations.

Review problems

1   Use a matrix to represent the coefficients of x in the following sets of linear equations:

(a) $2x_1 + 3x_2 + x_3 = y_1, \; x_1 - 2x_2 + 2x_3 = y_2, \; 3x_1 + x_2 = y_3,$

(b) $3x_1 + 2x_2 = 0, \; x_1 + x_2 = 4,$

(c) $x_1 + 2x_2 + x_3 = 1, \; x_2 + 5x_3 = 1, \; x_1 - 2x_2 + x_3 = 4$

3.2 Terminology

A *matrix* is a rectangular array of numbers. The numbers in the array are called *entries* in the matrix. The term *row* is used for a horizontal line of numbers and *column* for a vertical line of numbers. The *size* of a matrix is specified in terms of the number of rows and columns it contains. An array of numbers with p rows and q columns is called a $p \times q$ matrix. Thus, for example, a 2×2 matrix has two rows and two columns, e.g.

$$\begin{bmatrix} 1 & 2 \\ 3 & 4 \end{bmatrix}$$

A 3×2 matrix has three rows and two columns, e.g.

$$\begin{bmatrix} 1 & 2 \\ 3 & 4 \\ 5 & 6 \end{bmatrix}$$

A 2×1 matrix has two rows and one column, e.g.

$$\begin{bmatrix} 1 \\ 3 \end{bmatrix}$$

A matrix having just a single column but more than one row is called a *column matrix*. The 2×1 matrix illustrated above is a column matrix. Often the term *column vector* is used. A matrix having just a single row but more than one column is called a *row matrix*, e.g. the 1×2 matrix shown below. The term *row vector* is often used.

$$\begin{bmatrix} 1 & 2 \end{bmatrix}$$

A *square matrix* has the same number of rows and columns, as the 2×2 matrix illustrated earlier. The term *order* is used with square matrices to indicate the number of rows or the number of columns. Thus the 2×2 matrix is second order. A 3×3 matrix is third order. The term *diagonal matrix* is used for a square matrix with all entries zero, except those for which the row number equals the column number. We can put this another way: all the entries other than those on the leading diagonal are zero, the leading diagonal being the line going from the top left corner of the matrix to the bottom right corner. For example

$$\begin{bmatrix} 2 & 0 & 0 \\ 0 & 1 & 0 \\ 0 & 0 & 3 \end{bmatrix}$$

is a diagonal matrix. The term *unit matrix* or *identity matrix*, symbol **I**, is used for a diagonal matrix with all the entries on the leading diagonal having the value 1, e.g.

$$\begin{bmatrix} 1 & 0 & 0 \\ 0 & 1 & 0 \\ 0 & 0 & 1 \end{bmatrix}$$

The term *null matrix*, symbol **0**, is used for a matrix in which all the entries are 0, e.g.

$$\begin{bmatrix} 0 & 0 \\ 0 & 0 \end{bmatrix}$$

Review problems

2 For the following, identify the types of matrices concerned:

$$\text{(a)} \begin{bmatrix} 1 & 2 \\ -1 & 3 \end{bmatrix}, \text{(b)} \begin{bmatrix} 1 \\ 4 \\ 0 \end{bmatrix}, \text{(c)} \begin{bmatrix} 2 & 7 \end{bmatrix}, \text{(d)} \begin{bmatrix} 1 & 0 \\ 0 & 1 \end{bmatrix},$$

$$\text{(e)} \begin{bmatrix} 2 & 0 & 0 \\ 0 & 4 & 0 \\ 0 & 0 & 2 \end{bmatrix}, \text{(f)} \begin{bmatrix} 0 & 0 & 0 \\ 0 & 0 & 0 \\ 0 & 0 & 0 \end{bmatrix}, \text{(g)} \begin{bmatrix} 1 & 7 \\ 0 & -1 \\ 2 & 4 \end{bmatrix}$$

3.2.1 Equality of matrices

Two matrices are said to be *equal* if they have the same numbers of rows and columns and the corresponding entries in the two matrices are equal. Thus

$$\begin{bmatrix} 1 & 2 \\ 3 & 4 \end{bmatrix} \text{ and } \begin{bmatrix} 1 & 2 & 0 \\ 3 & 4 & 0 \end{bmatrix}$$

are *not* equal because there is an extra column in the second matrix.

$$\begin{bmatrix} 1 & 2 \\ 3 & 4 \end{bmatrix} \text{ and } \begin{bmatrix} 1 & 2 \\ 3 & 0 \end{bmatrix}$$

are *not* equal because, though they have the same numbers of rows and columns the corresponding entries are not equal.

3.2.2 Notation for matrices

The normal practice is to use capital letters in **bold** type to denote matrices, with lower case letters, not in bold type, used to denote numerical quantities inside matrices. Thus we might have

$$\mathbf{A} = \begin{bmatrix} a & b \\ c & d \\ e & f \end{bmatrix}$$

The a, b, c, etc. can be considered to be the coefficients of the corresponding linear equations (see chapter 2). In many texts column and row vectors, i.e. matrices consisting of a single row or a single column, are distinguished from other matrices by being represented by lower case letters in bold type, rather than capital letters. Thus, for example,

$$\mathbf{a} = \begin{bmatrix} a \\ b \\ c \end{bmatrix} \text{ or } \mathbf{a} = \begin{bmatrix} a & b & c \end{bmatrix}$$

Suffix notation is used to refer to particular elements in a matrix, the suffixes indicating the row followed by the column. The following matrix illustrates this:

$$\begin{bmatrix} a_{11} & a_{12} & a_{13} & \dots \\ a_{21} & a_{22} & a_{23} & \dots \\ a_{31} & a_{32} & a_{33} & \dots \\ \vdots & \vdots & \vdots & \end{bmatrix}$$

3.2.3 The transpose of a matrix

If $\mathbf{A}$ is a $p \times q$ matrix, then a related matrix is the *transpose* of $\mathbf{A}$ in which the rows and columns of $\mathbf{A}$ are interchanged. The result is thus a $q \times p$ matrix. The transpose is denoted by $\mathbf{A}^T$. Thus, for example, for the matrix

$$\mathbf{A} = \begin{bmatrix} 1 & 2 \\ 3 & 4 \end{bmatrix}$$

the transpose is

$$\mathbf{A}^T = \begin{bmatrix} 1 & 3 \\ 2 & 4 \end{bmatrix}$$

The first row has become the first column in the transpose, the second row the second column in the transpose. The first column has become the first row in the transpose, the second column the second row in the transpose.

If a square matrix $\mathbf{A}$ and its transpose $\mathbf{A}^T$ are identical then $\mathbf{A}$ is said to be a *symmetric* matrix. For example,

$$\mathbf{A} = \begin{bmatrix} 1 & 2 & 3 \\ 2 & 5 & 8 \\ 3 & 8 & 9 \end{bmatrix}$$

has a transpose of

$$\mathbf{A}^T = \begin{bmatrix} 1 & 2 & 3 \\ 2 & 5 & 8 \\ 3 & 8 & 9 \end{bmatrix}$$

Thus $\mathbf{A} = \mathbf{A}^T$. Note that a symmetric matrix is symmetrical about its leading diagonal (in the example this is 1 5 9).

If a square matrix $\mathbf{A}$ is such that the transpose $\mathbf{A}^T = -\mathbf{A}$, then $\mathbf{A}$ is said to be *skew symmetric*. For example,

$$\mathbf{A} = \begin{bmatrix} 0 & 2 \\ -2 & 0 \end{bmatrix}$$

has the transpose

$$\mathbf{A}^T = \begin{bmatrix} 0 & -2 \\ 2 & 0 \end{bmatrix}$$

The transpose is identical with the original matrix if the signs of the entries are reversed. It is skew symmetric.

Review problems

3 Determine the transpose of each of the following matrices:

$$\text{(a) } \begin{bmatrix} 1 & 4 \\ 2 & 0 \end{bmatrix}, \text{ (b) } \begin{bmatrix} 1 & 5 \\ 3 & 0 \\ -1 & 2 \end{bmatrix}, \text{ (c) } \begin{bmatrix} 1 & 2 & 5 \\ 4 & 2 & 1 \end{bmatrix}$$

4 Which of the following matrices are symmetric?

$$\mathbf{A} = \begin{bmatrix} 1 & 2 \\ 2 & 3 \end{bmatrix}, \quad \mathbf{B} = \begin{bmatrix} 2 & 5 \\ 5 & 2 \end{bmatrix}, \quad \mathbf{C} = \begin{bmatrix} 1 & 4 \\ 1 & 4 \end{bmatrix}$$

3.3 Addition and subtraction of matrices

Matrices of different sizes cannot be added or subtracted. Addition or subtraction of equal size matrices just involves the addition, or subtraction, of corresponding entries. For example, if we have the matrices

$$\mathbf{A} = \begin{bmatrix} 1 & 3 & 0 \\ 4 & 2 & 5 \end{bmatrix}$$

$$\mathbf{B} = \begin{bmatrix} -2 & 0 & 1 \\ 2 & -3 & 0 \end{bmatrix}$$

then

$$\mathbf{A} + \mathbf{B} = \begin{bmatrix} 1-2 & 3+0 & 0+1 \\ 4+2 & 2-3 & 5+0 \end{bmatrix}$$

and so

$$\mathbf{A} + \mathbf{B} = \begin{bmatrix} -1 & 3 & 1 \\ 6 & -1 & 5 \end{bmatrix}$$

Subtracting the matrices gives

$$\mathbf{A} - \mathbf{B} = \begin{bmatrix} 1+2 & 3-0 & 0-1 \\ 4-2 & 2+3 & 5-0 \end{bmatrix}$$

and so

$$\mathbf{A} - \mathbf{B} = \begin{bmatrix} 3 & 3 & -1 \\ 2 & 5 & 5 \end{bmatrix}$$

We can explain the above operations in terms of the coefficients of the simultaneous equations represented by the matrices. Thus, for the above matrices **A** and **B** we might have, for **A**,

$$x_1 + 3x_2 + 0x_3$$

and

$$4x_1 + 2x_2 + 5x_3$$

and for **B**,

$$-2x_1 + 0x_2 + x_3$$

and

$$2x_1 - 3x_2 + 0x_3$$

A + **B** thus represents, when we add the equations for the first rows,

$$(1 - 2)x_1 + (3 + 0)x_2 + (0 + 1)x_3$$

and, when we add the equations for the second rows,

$$(4 + 2)x_1 + (2 - 3)x_2 + (5 + 0)x_3$$

Review problems

5   For the following matrices, obtain where possible the sums and differences:

(a) **A** + **B**, (b) **A** − **B**, (c) **A** + **C**, (d) **C** + **D**, (e) **D** − **C**,

(f) **A** + **D**, (g) **B** − **A**, (h) **B** + **A**

$$\mathbf{A} = \begin{bmatrix} 1 & 2 \\ 3 & 4 \end{bmatrix}, \ \mathbf{B} = \begin{bmatrix} 2 & -3 \\ 0 & 1 \end{bmatrix},$$

$$\mathbf{C} = \begin{bmatrix} 1 & 2 & 3 \\ 4 & 5 & 6 \end{bmatrix}, \ \mathbf{D} = \begin{bmatrix} -1 & 0 & 1 \\ 2 & -3 & 0 \end{bmatrix}$$

6   The following spreadsheets show the costs incurred by a production department in the weeks of two months. Write the matrix for the costs for (a) each month, (b) the sum of the two months, (c) the change in costs from January to February.

January

	Week 1	Week 2	Week 3	Week 4
Materials	1000	600	800	1200
Power	400	300	400	500
Administration	100	100	120	120
Salaries	1200	1200	1300	1500

February

	Week 1	Week 2	Week 3	Week 4
Materials	1200	1000	800	800
Power	500	600	400	500
Administration	120	120	110	130
Salaries	1500	1700	1800	1600

3.3.1 Associative and commutative laws

The usual rules of arithmetic are followed in the addition, or subtraction, of matrices. Thus

$$A + B = B + A \qquad [1]$$

This is called the *commutative law*. We also have

$$(A + B) + C = A + (B + C) \qquad [2]$$

This is called the *associative law*. What these laws state is that the sequence in which we add matrices does not affect the result of the addition.

Review problems

7 By considering two specific examples of 2×2 matrices, show that $(A + B)^T = A^T + B^T$.

3.4 Multiplication by a constant

If we add two identical matrices, then the result is a matrix in which all the entries have twice the value. For example, if we have

$$A = \begin{bmatrix} 2 & 1 & 2 \\ 3 & 1 & 4 \\ 1 & 2 & 1 \end{bmatrix}$$

then $A + A$, i.e. $2A$, is

$$A + A = \begin{bmatrix} 2 & 1 & 2 \\ 3 & 1 & 4 \\ 1 & 2 & 1 \end{bmatrix} + \begin{bmatrix} 2 & 1 & 2 \\ 3 & 1 & 4 \\ 1 & 2 & 1 \end{bmatrix} = \begin{bmatrix} 4 & 2 & 4 \\ 6 & 2 & 8 \\ 2 & 4 & 2 \end{bmatrix}$$

Each entry is multiplied by 2. If we had $A + A + A$, i.e. $3A$, then we would have each entry multiplied by three. Thus we can state that the *product* of a constant c and a matrix A is the matrix obtained by multiplying each entry by c. The term *scalar* is often used for the constant c.

It can similarly be shown that

$$c(A + B) = cA + cB \qquad [3]$$

This is called the *distributive law*.

Example

For the following matrices, determine (a) $2\mathbf{A} + \mathbf{B}$, (b) $3\mathbf{A} - 2\mathbf{B}$.

$$\mathbf{A} = \begin{bmatrix} 1 & 2 \\ 2 & 5 \end{bmatrix}, \quad \mathbf{B} = \begin{bmatrix} 2 & -1 \\ 1 & 2 \end{bmatrix}$$

(a) To obtain $2\mathbf{A}$ we just multiply all the entries in matrix $\mathbf{A}$ by 2. Hence

$$2\mathbf{A} + \mathbf{B} = \begin{bmatrix} 2 & 4 \\ 4 & 10 \end{bmatrix} + \begin{bmatrix} 2 & -1 \\ 1 & 2 \end{bmatrix} = \begin{bmatrix} 4 & 3 \\ 5 & 12 \end{bmatrix}.$$

(b) To obtain $3\mathbf{A}$ we just multiply all the entries in matrix $\mathbf{A}$ by 3. To obtain $2\mathbf{B}$ we multiply all the entries in matrix $\mathbf{B}$ by 2. Hence

$$3\mathbf{A} - 2\mathbf{B} = \begin{bmatrix} 3 & 6 \\ 6 & 15 \end{bmatrix} - \begin{bmatrix} 4 & -2 \\ 2 & 4 \end{bmatrix} = \begin{bmatrix} -1 & 8 \\ 4 & 11 \end{bmatrix}$$

Example

Simplify the following matrix by removing a factor from it.

$$\mathbf{A} = \begin{bmatrix} 3 & 6 \\ 9 & 15 \end{bmatrix}$$

We can remove a factor of 3 from each of the entries to give

$$\mathbf{A} = 3 \begin{bmatrix} 1 & 2 \\ 3 & 5 \end{bmatrix}$$

Review problems

8 If the matrix $\mathbf{A}$ is given by

$$\mathbf{A} = \begin{bmatrix} 1 & 2 \\ 2 & 3 \\ 0 & 4 \end{bmatrix}$$

what are (a) $2\mathbf{A}$, (b) $3\mathbf{A}$, (c) $-1\mathbf{A}$, (d) $\tfrac{1}{2}\mathbf{A}$?

9 A company makes three sizes of each of two products. The following table gives the output over a period of 1 hour. What would be the output over a period of 6 hours if the output were to continue at the same rate? Express the result as a matrix.

	Small	Medium	Large
Product X	24	12	3
Product Y	16	6	1

10 Determine (a) $2\mathbf{A} + \mathbf{B}$, (b) $\mathbf{A} + 3\mathbf{B}$, (c) $\mathbf{A} - 2\mathbf{B}$, (d) $\mathbf{B} - 2\mathbf{A}$, (e) $2(\mathbf{A} + \mathbf{B}) - \mathbf{A}$, for the following matrices:

$$\mathbf{A} = \begin{bmatrix} 1 & 2 \\ 0 & 3 \\ 3 & 1 \end{bmatrix}, \quad \mathbf{B} = \begin{bmatrix} 2 & 3 \\ 1 & 2 \\ -2 & 0 \end{bmatrix}$$

11 Simplify the following matrices by removing factors from them:

$$\text{(a)} \begin{bmatrix} 4 & 6 \\ 2 & 10 \end{bmatrix}, \quad \text{(b)} \begin{bmatrix} 3 & -3 & 6 \\ 9 & 6 & 12 \end{bmatrix},$$

$$\text{(c)} \begin{bmatrix} ab & b^2 & b \\ b(a-2) & 3b & b^3 \\ ab^2 & b & b \end{bmatrix}$$

3.5 Multiplication of matrices

Consider two simultaneous equations

$$a_{11}x_1 + a_{12}x_2 = c_1 \quad \text{and} \quad a_{21}x_1 + a_{22}x_2 = c_2$$

in which x_1 and x_2 are two variables and the a and c terms are constants. To simplify these equations we introduce three matrices. The first matrix contains the coefficients, the second the variables and the third the constant terms on the right-hand sides of the equals signs. Thus we have

$$\mathbf{A} = \begin{bmatrix} a_{11} & a_{12} \\ a_{21} & a_{22} \end{bmatrix}, \quad \mathbf{x} = \begin{bmatrix} x_1 \\ x_2 \end{bmatrix}, \quad \mathbf{c} = \begin{bmatrix} c_1 \\ c_2 \end{bmatrix}$$

We define multiplication of matrices so that

$$\mathbf{Ax} = \mathbf{c} \tag{4}$$

which represents the two simultaneous equations we started with. Consider the conditions necessary for this to occur with

$$\begin{bmatrix} a_{11} & a_{12} \\ a_{21} & a_{22} \end{bmatrix} \begin{bmatrix} x_1 \\ x_2 \end{bmatrix} = \begin{bmatrix} c_1 \\ c_2 \end{bmatrix}$$

If we multiply the elements of the first row in the $\mathbf{A}$ matrix by the corresponding elements in a column in the $\mathbf{x}$ matrix, i.e. a_{11} by x_1 and a_{12} by x_2, and then the elements of the second row by the corresponding elements in the column of the $\mathbf{x}$ matrix, i.e. a_{21} by x_1 and a_{22} by x_2, we can obtain

$$\begin{bmatrix} a_{11}x_1 + a_{12}x_2 \\ a_{21}x_1 + a_{22}x_2 \end{bmatrix} = \begin{bmatrix} c_1 \\ c_2 \end{bmatrix}$$

The conditions for these two matrices to be equal are that

$$a_{11}x_1 + a_{12}x_2 = c_1 \quad \text{and} \quad a_{21}x_1 + a_{22}x_2 = c_2$$

Thus adopting this procedure for multiplication gives the simultaneous equations.

With a 2×2 matrix $\mathbf{A}$ multiplied by another 2×2 matrix $\mathbf{B}$, e.g.

$$\begin{bmatrix} a_{11} & a_{12} \\ a_{21} & a_{22} \end{bmatrix} \begin{bmatrix} b_{11} & b_{12} \\ b_{21} & b_{22} \end{bmatrix} = \begin{bmatrix} c_{11} & c_{12} \\ c_{21} & c_{22} \end{bmatrix}$$

we multiply the first row of $\mathbf{A}$ by the corresponding elements in the first column of the $\mathbf{B}$ matrix to obtain the entries c_{11} in the product $\mathbf{C}$ matrix. Then we multiply the second row of $\mathbf{A}$ by the corresponding elements in the first column of the $\mathbf{B}$ matrix to obtain the entry c_{21} in the product matrix.

$$\begin{bmatrix} a_{11} & a_{12} \\ a_{21} & a_{22} \end{bmatrix} \begin{bmatrix} b_{11} & \cdots \\ b_{21} & \cdots \end{bmatrix} = \begin{bmatrix} c_{11} & \cdots \\ c_{21} & \cdots \end{bmatrix}$$

We then repeat this procedure with the entries in the $\mathbf{A}$ matrix multiplying the entries in the second column of the $\mathbf{B}$ matrix to give the c_{12} and c_{22} entries.

$$\begin{bmatrix} a_{11} & a_{12} \\ a_{21} & a_{22} \end{bmatrix} \begin{bmatrix} \cdots & b_{12} \\ \cdots & b_{22} \end{bmatrix} = \begin{bmatrix} \cdots & c_{12} \\ \cdots & c_{22} \end{bmatrix}$$

Thus the result is

$$\begin{bmatrix} a_{11}b_{11} + a_{12}b_{21} & a_{11}b_{12} + a_{12}b_{22} \\ a_{21}b_{11} + a_{22}b_{21} & a_{11}b_{12} + a_{12}b_{22} \end{bmatrix} = \begin{bmatrix} c_{11} & c_{12} \\ c_{21} & c_{22} \end{bmatrix}$$

For example, for

$$\mathbf{A} = \begin{bmatrix} 1 & 2 \\ 3 & 4 \end{bmatrix} \quad \text{and} \quad \mathbf{B} = \begin{bmatrix} 5 & 6 \\ 7 & 8 \end{bmatrix}$$

the product **AB** is

$$\begin{bmatrix} 1 & 2 \\ 3 & 4 \end{bmatrix}\begin{bmatrix} 5 & 6 \\ 7 & 8 \end{bmatrix} = \begin{bmatrix} 1\times5+2\times7 & 1\times6+2\times8 \\ 3\times5+4\times7 & 3\times6+4\times8 \end{bmatrix}$$

$$= \begin{bmatrix} 19 & 22 \\ 43 & 50 \end{bmatrix}$$

For multiplication of two matrices **A** and **B** to be possible we *must* have the number of columns in the first matrix equal to the number of rows in the second matrix. The matrix resulting from the multiplication will have entries of c_{ij} equal to the product of row i in matrix **A** by column j in matrix **B**. A way of representing this that you might find useful is as follows:

$$\begin{array}{c} \text{column} \\ j \end{array}$$

$$\begin{bmatrix} b_{1j} \\ b_{2j} \\ \vdots \end{bmatrix} \text{Matrix } \mathbf{B}$$

$$\text{row } i \begin{bmatrix} a_{i1} & a_{i2} & \cdots \end{bmatrix}\begin{bmatrix} c_{ij} \end{bmatrix} \qquad [5]$$

$$\begin{array}{cc} \text{Matrix } \mathbf{A} & \text{Product} \end{array}$$

Each entry c_{ij} of the product in the above representation occupies the intersection of the row of the first matrix **A** with the row of the second matrix **B**, these being the row and column used to compute it. Thus for the example considered earlier, we can write

$$\begin{bmatrix} 5 & 6 \\ 7 & 8 \end{bmatrix}\text{Matrix } \mathbf{B}$$

$$\begin{bmatrix} 1 & 2 \\ 3 & 4 \end{bmatrix}\begin{bmatrix} 19 & 22 \\ 43 & 50 \end{bmatrix}$$

$$\begin{array}{cc} \text{Matrix } \mathbf{A} & \text{Product} \end{array}$$

With matrix multiplication

$$\mathbf{AB} \neq \mathbf{BA} \qquad [6]$$

Matrix multiplication is said to be *not commutative*. Thus, for the product of the two 2×2 matrices given above, the product **BA** is

$$\begin{bmatrix} 5 & 6 \\ 7 & 8 \end{bmatrix} \begin{bmatrix} 1 & 2 \\ 3 & 4 \end{bmatrix} = \begin{bmatrix} 5\times1+6\times3 & 5\times2+6\times4 \\ 7\times1+8\times3 & 7\times2+8\times4 \end{bmatrix}$$

$$= \begin{bmatrix} 23 & 34 \\ 31 & 46 \end{bmatrix}$$

We thus have **AB** $\neq$ **BA**, i.e.

$$\begin{bmatrix} 19 & 22 \\ 43 & 50 \end{bmatrix} \neq \begin{bmatrix} 23 & 34 \\ 31 & 46 \end{bmatrix}$$

A further difference from normal algebra is that we can have the product **AB** = **0**, the **0** being a matrix with all entries 0, without **A** or **B** being zero. For example

$$\begin{bmatrix} 1 & 2 & 4 \\ 2 & 4 & 8 \end{bmatrix} \begin{bmatrix} 2 \\ 3 \\ -2 \end{bmatrix} = \begin{bmatrix} 1\times2+2\times3+4\times(-2) \\ 2\times2+4\times3+8\times(-2) \end{bmatrix} = \begin{bmatrix} 0 \\ 0 \end{bmatrix}$$

The *associative* and *distributive* laws do hold for matrix multiplication, i.e.

$$(AB)C = A(CB) \tag{7}$$

$$A(B + C) = AB + AC \tag{8}$$

$$(A + B)C = AC + BC \tag{9}$$

Unit matrices, symbol **I**, have ones on the main diagonal and zeros for all other entries. Multiplication of a square matrix $n \times n$ by a suitable unit matrix, i.e. also $n \times n$, plays the same role in matrix algebra as multiplying by 1 in conventional arithmetic, i.e.

$$AI = IA = A \tag{10}$$

For example,

$$\begin{bmatrix} 2 & 3 & 1 \\ 1 & -1 & 2 \\ 2 & 4 & 2 \end{bmatrix} \begin{bmatrix} 1 & 0 & 0 \\ 0 & 1 & 0 \\ 0 & 0 & 1 \end{bmatrix} = \begin{bmatrix} 2 & 3 & 1 \\ 1 & -1 & 2 \\ 2 & 4 & 2 \end{bmatrix}$$

$$\begin{bmatrix} 1 & 0 & 0 \\ 0 & 1 & 0 \\ 0 & 0 & 1 \end{bmatrix} \begin{bmatrix} 2 & 3 & 1 \\ 1 & -1 & 2 \\ 2 & 4 & 2 \end{bmatrix} = \begin{bmatrix} 2 & 3 & 1 \\ 1 & -1 & 2 \\ 2 & 4 & 2 \end{bmatrix}$$

Example

Determine the product of the matrices **A** and **B**, where

$$\mathbf{A} = \begin{bmatrix} 1 & 2 & 1 \\ -1 & 0 & 3 \end{bmatrix}, \quad \mathbf{B} = \begin{bmatrix} 2 \\ -2 \\ 1 \end{bmatrix}$$

Matrix **A** has the same number of columns as matrix **B** has rows, so we can multiply them. Thus

$$\begin{bmatrix} 1 & 2 & 1 \\ -1 & 0 & 3 \end{bmatrix} \begin{bmatrix} 2 \\ -2 \\ 1 \end{bmatrix} = \begin{bmatrix} 1 \times 2 + 2 \times (-2) + 1 \times 1 \\ (-1) \times 2 + 0 \times (-2) + 3 \times 1 \end{bmatrix}$$

$$= \begin{bmatrix} -1 \\ 1 \end{bmatrix}$$

Example

Determine the product of the matrices **A** and **B**, where

$$\mathbf{A} = \begin{bmatrix} 1 & 0 \\ 2 & 3 \end{bmatrix}, \quad \mathbf{B} = \begin{bmatrix} 2 & 1 \\ -1 & 2 \end{bmatrix}$$

Matrix **A** has the same number of columns as matrix **B** has rows, so we can multiply them. Thus

$$\mathbf{AB} = \begin{bmatrix} 1 \times 2 + 0 \times (-1) & 1 \times 1 + 0 \times 2 \\ 2 \times 2 + 3(-1) & 2 \times 1 + 3 \times 2 \end{bmatrix} = \begin{bmatrix} 2 & 1 \\ 1 & 8 \end{bmatrix}$$

Example

Determine the product of the matrices **A** and **B**, where

$$\mathbf{A} = \begin{bmatrix} 2 & 1 & -1 \\ 1 & 4 & 2 \\ 1 & -3 & 1 \end{bmatrix}, \quad \mathbf{B} = \begin{bmatrix} 1 & 0 & 1 \\ 0 & 1 & 0 \\ 0 & 0 & 1 \end{bmatrix}$$

Matrix **A** has the same number of columns as matrix **B** has rows, so we can multiply them. Thus

$$\mathbf{AB} = \begin{bmatrix} 2 & 1 & 2-1 \\ 1 & 4 & 1+2 \\ 1 & -3 & 1+1 \end{bmatrix} = \begin{bmatrix} 2 & 1 & 1 \\ 1 & 4 & 3 \\ 1 & -3 & 2 \end{bmatrix}$$

Review problems

12 Determine, if possible, the products **AB**, **BA**, **Ac** and **Bc**, where

(a) $\mathbf{A} = \begin{bmatrix} 1 & 0 & 2 \\ -1 & 1 & 3 \end{bmatrix}$, $\mathbf{B} = \begin{bmatrix} 2 & 1 \\ -2 & 0 \\ 4 & 5 \end{bmatrix}$, $\mathbf{c} = \begin{bmatrix} 1 \\ 2 \\ 3 \end{bmatrix}$

(b) $\mathbf{A} = \begin{bmatrix} 6 & 1 & 3 \\ -1 & 1 & 2 \\ 4 & 1 & 3 \end{bmatrix}$, $\mathbf{B} = \begin{bmatrix} 1 & 5 & 2 \\ -1 & 0 & 1 \\ 3 & 2 & 4 \end{bmatrix}$, $\mathbf{c} = \begin{bmatrix} 2 \\ 1 \\ 1 \end{bmatrix}$

(c) $\mathbf{A} = \begin{bmatrix} 1 & -1 & 2 \\ 1 & 3 & 4 \\ 2 & 5 & -1 \end{bmatrix}$, $\mathbf{B} = \begin{bmatrix} 1 & 0 \\ 1 & 3 \\ 1 & 1 \end{bmatrix}$, $\mathbf{c} = \begin{bmatrix} 1 & 2 & 1 \end{bmatrix}$

13 For the following matrices, determine $\mathbf{A}^{\mathrm{T}}\mathbf{b}$:

$\mathbf{A} = \begin{bmatrix} 1 & 1 & 0 \\ 2 & 0 & 1 \end{bmatrix}$, $\mathbf{b} = \begin{bmatrix} -1 \\ 2 \end{bmatrix}$

14 Determine the product of the following matrices:

$\begin{bmatrix} 1 & 4 & 1 \\ 2 & -1 & 3 \end{bmatrix} \begin{bmatrix} x \\ y \\ z \end{bmatrix}$

15 A production department makes three models of a particular product, these being small, medium and large sizes. Three subassemblies are used, in different numbers, in the three models. The following table shows the numbers used:

	Small	Medium	Large
Subassembly A	2	4	9
Subassembly B	1	3	5
Subassembly C	2	5	8

Represent the data as a matrix and then use matrix multiplication to determine the numbers of each subassembly required to make 6 small, 12 medium and 5 large models.

3.6 Inversion of a matrix

At the beginning of section 3.5 we considered two simultaneous equations

$$a_{11}x_1 + a_{12}x_2 = c_1 \text{ and } a_{21}x_1 + a_{22}x_2 = c_2$$

in which x_1 and x_2 are two variables and the a and c terms are constants. To simplify these equations three matrices **A**, **x** and **c** were introduced with the matrix **A** containing the coefficients, the matrix **x** the variables and the matrix **c** the constant terms on the right-hand sides of the equals signs. We then had

$$\mathbf{Ax} = \mathbf{c} \qquad\qquad [11]$$

If we had conventional numbers, e.g. $ax = c$, then we could obtain x by $x = c/a$. Unfortunately we cannot do this with matrices since division does not exist with matrices. With conventional numbers we can, however, write the equation as $x = a^{-1}c$, where a^{-1} is the reciprocal or inverse of a. What we have effectively done is multiply both sides of the equation $ax = c$ by a^{-1} to obtain $a^{-1}ax = a^{-1}c$. We then utilise the fact that $a \times a^{-1} = 1$ and so obtain $x = a^{-1}c$.

With matrices we adopt a similar procedure. Multiplying both sides of equation [11] by the inverse of matrix **A**, denoted by $\mathbf{A}^{-1}$, we obtain

$$\mathbf{A}^{-1}\mathbf{Ax} = \mathbf{A}^{-1}\mathbf{c}$$

In a similar way to conventional numbers we define the inverse matrix by the relationship

$$\mathbf{AA}^{-1} = \mathbf{I} \qquad\qquad [12]$$

where **I** is the unit matrix. Thus

$$\mathbf{x} = \mathbf{A}^{-1}\mathbf{c} \qquad\qquad [13]$$

To obtain the **x** matrix, all we need to do is multiply the **c** matrix by the inverse **A** matrix.

3.6.1 Inverse of a 2 × 2 matrix

Consider the matrix

$$\mathbf{A} = \begin{bmatrix} a & b \\ c & d \end{bmatrix} \qquad\qquad [14]$$

Then with

$$I = \begin{bmatrix} 1 & 0 \\ 0 & 1 \end{bmatrix}$$

we have to find what we need to multiply **A** by to get **I**. Let $\mathbf{A}^{-1}$ be given by

$$\mathbf{A}^{-1} = \begin{bmatrix} p & q \\ r & s \end{bmatrix}$$

Then we must have

$$\begin{bmatrix} a & b \\ c & d \end{bmatrix} \begin{bmatrix} p & q \\ r & s \end{bmatrix} = \begin{bmatrix} 1 & 0 \\ 0 & 1 \end{bmatrix}$$

Then we require

$$ap + br = 1$$

$$cp + dr = 0$$

$$aq + bs = 0$$

$$cq + ds = 1$$

We can eliminate r from the first pair of equations and obtain

$$ap - \frac{bcp}{d} = 1$$

Hence

$$p = \frac{d}{ad - bc}$$

and

$$r = -\frac{cp}{d} = \frac{-c}{ad - bc}$$

Eliminating s from the second pair of equations gives

$$q = \frac{-b}{ad - bc}$$

and

$$s = -\frac{aq}{b} = \frac{a}{ad - bc}$$

Thus the inverse is

$$\mathbf{A}^{-1} = \begin{bmatrix} p & q \\ r & s \end{bmatrix} = \begin{bmatrix} \dfrac{d}{ad-bc} & \dfrac{-b}{ad-bc} \\ \dfrac{-c}{ad-bc} & \dfrac{a}{ad-bc} \end{bmatrix}$$

We can take the factor $1/(ad - bc)$ out of the matrix to give the inverse of the matrix $\mathbf{A}$ (equation [14]) as

$$\mathbf{A}^{-1} = \frac{1}{ad-bc} \begin{bmatrix} d & -b \\ -c & a \end{bmatrix} \qquad\qquad [15]$$

The expression $ad - bc$ is known as the *determinant* of the matrix (see chapter 4 for a discussion of determinants).

We can summarise this result as:

1   In the matrix, a and d are swapped over.
2   In the matrix, the signs of b and c are changed.
3   Divide the resulting matrix by $(ad - bc)$.

A matrix does not always have an inverse. For example,

$$\mathbf{A} = \begin{bmatrix} 1 & -1 \\ -1 & 1 \end{bmatrix}$$

has no inverse. The determinant, i.e. $ad - bc$, is equal to 0, thus the 1/0 that occurs means there is no inverse. This means that the set of simultaneous linear equations from which the matrix was derived has no unique solution. There is no matrix which when multiplied by $\mathbf{A}$ gives $\mathbf{I}$. In general, a matrix $\mathbf{A}$ does not have an inverse if the determinant of $\mathbf{A}$ is zero.

Example

Determine the inverse of the matrix

$$\mathbf{A} = \begin{bmatrix} 1 & 2 \\ -1 & 5 \end{bmatrix}$$

Using equation [15],

$$\mathbf{A}^{-1} = \frac{1}{(1 \times 5) - (2 \times -1)} \begin{bmatrix} 5 & -2 \\ 1 & 1 \end{bmatrix} = \frac{1}{7} \begin{bmatrix} 5 & -2 \\ 1 & 1 \end{bmatrix}$$

The determinant of the matrix $\mathbf{A}$ has the value 7.

Review problems

16 Determine the inverses for the following matrices:

$$A = \begin{bmatrix} 1 & -1 \\ 1 & 1 \end{bmatrix}, \quad B = \begin{bmatrix} 3 & 4 \\ 2 & 5 \end{bmatrix}, \quad C = \begin{bmatrix} 2 & -3 \\ -4 & 5 \end{bmatrix},$$

$$D = \begin{bmatrix} -1 & 2 \\ 4 & 2 \end{bmatrix}$$

3.6.2 The inverse by row operations

To find the inverse A^{-1} of a matrix A we have to find a matrix which when multiplied by A gives I (equation [12]), i.e.

$$AA^{-1} = I$$

Thus, for example, if

$$A = \begin{bmatrix} a_{11} & a_{12} \\ a_{21} & a_{22} \end{bmatrix}$$

and we take the inverse to be

$$A^{-1} = \begin{bmatrix} b_{11} & b_{12} \\ b_{21} & b_{22} \end{bmatrix}$$

then we must have

$$\begin{bmatrix} a_{11} & a_{12} \\ a_{21} & a_{22} \end{bmatrix} \begin{bmatrix} b_{11} & b_{12} \\ b_{21} & b_{22} \end{bmatrix} = \begin{bmatrix} 1 & 0 \\ 0 & 1 \end{bmatrix}$$

and so

$$a_{11}b_{11} + a_{12}b_{21} = 1$$

$$a_{21}b_{11} + a_{22}b_{21} = 0$$

$$a_{11}b_{12} + a_{12}b_{12} = 0$$

$$a_{21}b_{21} + a_{22}b_{22} = 1$$

Because these are linear equations we can multiply an equation throughout by a constant, provided we do it to both sides of the equation. We can also add or subtract equations, or interchange the sequence of the equations. Whenever we do such operations we must do them equally to both sides of the equals sign. What we are doing is carrying out certain elementary operations on rows of the matrix **A** and **I**.

Thus a technique that we can use, provided the inverse exists, is to operate by row manipulation on the matrix **A** and transform it into the matrix **I**. If we do the same manipulations with **I** then the result is $\mathbf{A}^{-1}$. The basic row procedures that can be used are:

1   Multiply a row by a constant.
2   Add or subtract one row from another.
3   Interchange any two rows.

The following example illustrates the process. Another method involving the adjoint matrix is discussed in the next chapter.

Consider the matrix

$$\mathbf{A} = \begin{bmatrix} 2 & 4 & 6 \\ 3 & 1 & -2 \\ 4 & 5 & 6 \end{bmatrix}$$

To find the inverse we carry out row operations on this matrix and the corresponding **I** matrix.

$$\begin{bmatrix} 2 & 4 & 6 \\ 3 & 1 & -2 \\ 4 & 5 & 6 \end{bmatrix} \quad \begin{bmatrix} 1 & 0 & 0 \\ 0 & 1 & 0 \\ 0 & 0 & 1 \end{bmatrix}$$

Row 1
$\times 1/2$
$$\begin{bmatrix} 1 & 2 & 3 \\ 3 & 1 & -2 \\ 4 & 5 & 6 \end{bmatrix} \quad \begin{bmatrix} \frac{1}{2} & 0 & 0 \\ 0 & 1 & 0 \\ 0 & 0 & 1 \end{bmatrix}$$

Row 2
$\times 1/3$
$$\begin{bmatrix} 1 & 2 & 3 \\ 1 & \frac{1}{3} & -\frac{2}{3} \\ 4 & 5 & 6 \end{bmatrix} \quad \begin{bmatrix} \frac{1}{2} & 0 & 0 \\ 0 & \frac{1}{3} & 0 \\ 0 & 0 & 1 \end{bmatrix}$$

Row 3
$\times 1/4$
$$\begin{bmatrix} 1 & 2 & 3 \\ 1 & \frac{1}{3} & -\frac{2}{3} \\ 1 & \frac{5}{4} & \frac{3}{2} \end{bmatrix} \quad \begin{bmatrix} \frac{1}{2} & 0 & 0 \\ 0 & \frac{1}{3} & 0 \\ 0 & 0 & \frac{1}{4} \end{bmatrix}$$

Row 2
− Row 1
$$\begin{bmatrix} 1 & 2 & 3 \\ 0 & -\frac{5}{3} & -\frac{11}{3} \\ 1 & \frac{5}{4} & \frac{3}{2} \end{bmatrix} \quad \begin{bmatrix} \frac{1}{2} & 0 & 0 \\ -\frac{1}{2} & \frac{1}{3} & 0 \\ 0 & 0 & \frac{1}{4} \end{bmatrix}$$

Row 3
− Row 1
$$\begin{bmatrix} 1 & 2 & 3 \\ 0 & -\frac{5}{3} & -\frac{11}{3} \\ 0 & -\frac{3}{4} & -\frac{3}{2} \end{bmatrix} \quad \begin{bmatrix} \frac{1}{2} & 0 & 0 \\ -\frac{1}{2} & \frac{1}{3} & 0 \\ -\frac{1}{2} & 0 & \frac{1}{4} \end{bmatrix}$$

Row 2
× −3/5
$$\begin{bmatrix} 1 & 2 & 3 \\ 0 & 1 & \frac{11}{5} \\ 0 & -\frac{3}{4} & -\frac{3}{2} \end{bmatrix} \quad \begin{bmatrix} \frac{1}{2} & 0 & 0 \\ \frac{3}{10} & -\frac{1}{5} & 0 \\ -\frac{1}{2} & 0 & \frac{1}{4} \end{bmatrix}$$

Row 3 +
3/4 Row 2
$$\begin{bmatrix} 1 & 2 & 3 \\ 0 & 1 & \frac{11}{5} \\ 0 & 0 & \frac{3}{20} \end{bmatrix} \quad \begin{bmatrix} \frac{1}{2} & 0 & 0 \\ \frac{3}{10} & -\frac{1}{5} & 0 \\ -\frac{11}{40} & -\frac{3}{20} & \frac{1}{4} \end{bmatrix}$$

Row 3
× 20/3
$$\begin{bmatrix} 1 & 2 & 3 \\ 0 & 1 & \frac{11}{5} \\ 0 & 0 & 1 \end{bmatrix} \quad \begin{bmatrix} \frac{1}{2} & 0 & 0 \\ \frac{3}{10} & -\frac{1}{5} & 0 \\ -\frac{11}{6} & -1 & \frac{5}{3} \end{bmatrix}$$

Row 2 −
11/5 Row 3
$$\begin{bmatrix} 1 & 2 & 3 \\ 0 & 1 & 0 \\ 0 & 0 & 1 \end{bmatrix} \quad \begin{bmatrix} \frac{1}{2} & 0 & 0 \\ \frac{13}{3} & 2 & -\frac{11}{3} \\ -\frac{11}{6} & -1 & \frac{5}{3} \end{bmatrix}$$

Row 1
− 2 Row 2
− 3 Row 3
$$\begin{bmatrix} 1 & 0 & 0 \\ 0 & 1 & 0 \\ 0 & 0 & 1 \end{bmatrix} \quad \begin{bmatrix} -\frac{8}{3} & -1 & \frac{7}{3} \\ \frac{13}{3} & 2 & -\frac{11}{3} \\ -\frac{11}{6} & -1 & \frac{5}{3} \end{bmatrix}$$

Thus the inverse matrix can be written as

$$\mathbf{A}^{-1} = \frac{1}{6} \begin{bmatrix} -16 & -6 & 14 \\ 26 & 12 & -22 \\ -11 & -6 & 10 \end{bmatrix}$$

We can check this by multiplying $\mathbf{A}^{-1}$ by $\mathbf{A}$, when we should obtain $\mathbf{I}$.

If in the manipulation of rows we end up with a row of zeros, then the matrix has no inverse. There is then no unique solution to the set of simultaneous equations from which the matrix was derived.

Review problems

17 Use row operations to obtain the inverse matrices of the following:

$$\mathbf{A} = \begin{bmatrix} 1 & 0 & 2 \\ 0 & 1 & 2 \\ 1 & 2 & 0 \end{bmatrix}, \quad \mathbf{B} = \begin{bmatrix} 1 & -2 & 0 \\ 0 & 1 & 3 \\ 1 & 0 & -1 \end{bmatrix},$$

$$\mathbf{C} = \begin{bmatrix} 2 & 3 & -1 \\ 0 & 1 & 1 \\ -1 & 2 & 1 \end{bmatrix}, \quad \mathbf{D} = \begin{bmatrix} 1 & 0 & 1 \\ -1 & 1 & 1 \\ 0 & 1 & 0 \end{bmatrix},$$

$$\mathbf{E} = \begin{bmatrix} 1 & 1 & 1 \\ 1 & 2 & 3 \\ 0 & 1 & 1 \end{bmatrix}, \quad \mathbf{F} = \begin{bmatrix} 1 & 2 & -1 \\ 2 & 1 & 0 \\ -1 & 1 & 2 \end{bmatrix}$$

3.7 Simultaneous equations

The simultaneous equations

$$2x + y = 3$$

$$x + 4y = 1$$

can be written in matrix form as

$$\begin{bmatrix} 2 & 1 \\ 1 & 4 \end{bmatrix} \begin{bmatrix} x \\ y \end{bmatrix} = \begin{bmatrix} 3 \\ 1 \end{bmatrix}$$

We can solve this by using the inverse matrix (equation [15]),

$$\text{inverse of } \begin{bmatrix} 2 & 1 \\ 1 & 4 \end{bmatrix} = \frac{1}{8-1} \begin{bmatrix} 4 & -1 \\ -1 & 2 \end{bmatrix}$$

and so

$$\begin{bmatrix} x \\ y \end{bmatrix} = \frac{1}{7} \begin{bmatrix} 4 & -1 \\ -1 & 2 \end{bmatrix} \begin{bmatrix} 3 \\ 1 \end{bmatrix}$$

and $x = (12 - 1)/7 = 11/7$, $y = (-3 + 2)/7 = -1/7$.

The procedure for using matrices to solve simultaneous equations is thus:

1   Write each equation in the form $a_1 x + b_1 x + c_1 z + \ldots = d_1$.
2   Write the matrix equation corresponding to the equations, i.e.

$$\begin{bmatrix} a_1 & b_1 & c_1 & \cdots \\ \vdots & \vdots & \vdots & \end{bmatrix} \begin{bmatrix} x \\ y \\ z \\ \vdots \end{bmatrix} = \begin{bmatrix} d_1 \\ \vdots \end{bmatrix}$$

i.e. $\mathbf{Ax} = \mathbf{d}$.
3   Determine the inverse of the matrix $\mathbf{A}$, i.e. the inverse of

$$\begin{bmatrix} a_1 & b_1 & c_1 & \cdots \\ \vdots & \vdots & \vdots & \end{bmatrix}$$

This is $\mathbf{A}^{-1}$.

4 Multiply the inverse matrix by

$$\begin{bmatrix} d_1 \\ \vdots \end{bmatrix}$$

i.e. $\mathbf{x} = \mathbf{A}^{-1}\mathbf{d}$, to give the values of x, y, z, etc.

Review problems

18 By determining the inverse matrix, solve the following simultaneous equations:

(a) $x + 2y = 2$, $3x + 4y = 4$,

(b) $3x + 5y = 2$, $4x - 3y = 7$,

(c) $x - 2y = 3$, $x + 4y = 5$

19 By determining the inverse matrix, solve the following simultaneous equations:

(a) $x - z = 1$, $-x + y + z = 2$, $y = 3$,

(b) $x + 3y + 3z = 2$, $x + 4y + 3z = 3$, $x + 3y + 4z = 2$,

(c) $x - 2y = 4$, $y + 3z = 12$, $x - z = -2$,

(d) $x + y + z = 5$, $2x + y - z = 5$, $-x + z = -1$

Further problems

20 For the following matrices, determine (a) $\mathbf{A} + \mathbf{B}$, (b) $\mathbf{C} + \mathbf{D}$, (c) $\mathbf{A} - \mathbf{B}$, (d) $\mathbf{B} - \mathbf{C}$, (e) $\mathbf{A} + \mathbf{B} - \mathbf{D}$, (f) $\mathbf{D} - \mathbf{A} - \mathbf{B}$, (g) $2\mathbf{A}$, (h) $3\mathbf{B}$, (i) $2\mathbf{C} + \mathbf{D}$, (j) $\mathbf{A} - 2\mathbf{B}$, (k) $\mathbf{D} - 2\mathbf{A}$, (l) $\mathbf{A} + 0.5\mathbf{D}$.

$$\mathbf{A} = \begin{bmatrix} 2 & 1 & 0 \\ 1 & 2 & 1 \\ 2 & 2 & 0 \end{bmatrix}, \quad \mathbf{B} = \begin{bmatrix} 1 & 1 & 1 \\ 2 & 0 & 2 \\ -1 & 0 & 2 \end{bmatrix},$$

$$\mathbf{C} = \begin{bmatrix} -1 & 2 & 3 \\ 2 & 1 & 4 \\ -1 & -2 & 1 \end{bmatrix}, \quad \mathbf{D} = \begin{bmatrix} 0 & 0 & 2 \\ 1 & 1 & 4 \\ 2 & -1 & 1 \end{bmatrix}$$

21 Determine, if possible, for the following matrices: (a) **AB**, (b) **BA**, (c) **AC**, (d) **AD**, (e) **CD**.

$$A = \begin{bmatrix} 3 & 2 \\ 5 & 6 \end{bmatrix}, \; B = \begin{bmatrix} 1 & 3 \\ -1 & 1 \end{bmatrix}, \; C = \begin{bmatrix} 1 & 4 \end{bmatrix}, \; D = \begin{bmatrix} 1 \\ 4 \end{bmatrix}$$

22 Determine, if possible, for the following matrices: (a) **AB**, (b) **BA**, (c) **AC**, (d) **CA**.

$$A = \begin{bmatrix} 1 & 0 & 3 \\ 2 & 1 & 1 \\ 1 & 0 & 1 \end{bmatrix}, \; B = \begin{bmatrix} -1 \\ 2 \\ 1 \end{bmatrix}, \; C = \begin{bmatrix} 1 & 0 & 0 \\ 0 & 1 & 0 \\ 0 & 0 & 1 \end{bmatrix}$$

23 Determine the product of the following matrices:

$$\begin{bmatrix} 1 & 2 \\ -1 & 4 \end{bmatrix} \begin{bmatrix} x \\ y \end{bmatrix}$$

24 Determine the product of the following matrices:

$$\begin{bmatrix} a & 0 \\ 0 & b \end{bmatrix} \begin{bmatrix} 0 & a \\ b & 0 \end{bmatrix}$$

25 Determine the product **ABC** where

$$A = \begin{bmatrix} 2 & 1 \\ -1 & 0 \\ 2 & 3 \end{bmatrix}, \; B = \begin{bmatrix} 3 & 0 \\ -2 & 1 \end{bmatrix}, \; C = \begin{bmatrix} -1 & 2 & 3 \\ 4 & 0 & 1 \end{bmatrix}$$

26 For the following matrix, show that AA^{T} is a symmetric matrix.

$$A = \begin{bmatrix} a & b \\ c & d \end{bmatrix}$$

27 Determine the transposes of the following matrices:

$$A = \begin{bmatrix} 1 & 2 & 3 \\ 4 & 5 & 6 \end{bmatrix}, \; B = \begin{bmatrix} 1 & 2 & 3 \end{bmatrix}, \; C = \begin{bmatrix} 1 & 2 \\ 3 & 4 \\ 5 & 6 \end{bmatrix}$$

28 The following matrices **A** and **B** are symmetric. Is **A** + **B** symmetric?

$$\mathbf{A} = \begin{bmatrix} 1 & 2 & -3 \\ 2 & 2 & 0 \\ -3 & 0 & 3 \end{bmatrix}, \ \mathbf{B} = \begin{bmatrix} 3 & 1 & -2 \\ 1 & 1 & 4 \\ -2 & 4 & -1 \end{bmatrix}$$

29 Determine the inverse matrices for the following 2×2 matrices:

$$\mathbf{A} = \begin{bmatrix} 1 & 4 \\ -1 & 3 \end{bmatrix}, \ \mathbf{B} = \begin{bmatrix} 4 & -2 \\ 1 & 2 \end{bmatrix}, \ \mathbf{C} = \begin{bmatrix} 2 & 1 \\ 3 & 2 \end{bmatrix},$$

$$\mathbf{D} = \begin{bmatrix} 3 & 5 \\ 1 & 2 \end{bmatrix}$$

30 Use row operations to obtain the inverse matrices of the following:

$$\mathbf{A} = \begin{bmatrix} 3 & 2 & 0 \\ -1 & -2 & 4 \\ 2 & -1 & -3 \end{bmatrix}, \ \mathbf{B} = \begin{bmatrix} 4 & 2 & 0 \\ 0 & 1 & 1 \\ 3 & 2 & 0 \end{bmatrix},$$

$$\mathbf{C} = \begin{bmatrix} 1 & 0 & 1 \\ 0 & 1 & 1 \\ 1 & 1 & 0 \end{bmatrix}$$

31 Solve, by using inverse matrices, the following sets of simultaneous equations:

(a) $x + 2y = 1, \ 3x + 5y = 2,$

(b) $x + 3y = 2, \ x - y = 1,$

(c) $3x - 5y = 1, \ -2x + 3y = 2,$

(d) $2x + y + z = 1, \ x + 2y + 2z = 2, \ x + 3y + 2z = 1,$

(e) $3x + 2y + z = 1, \ 2y + 2z = 2, \ -z = 4,$

(f) $2x + y = 3, \ x + 2y - z = 2, \ -x + y + 2z = 4$

$\mathbf{4}$ Determinants

4.1 Determinants

Consider two simultaneous equations with two variables x_1 and x_2,

$$a_{11}x_1 + a_{12}x_2 = c_1 \text{ and } a_{21}x_1 + a_{22}x_2 = c_2$$

These can be represented by the matrix equation

$$\begin{bmatrix} a_{11} & a_{12} \\ a_{21} & a_{22} \end{bmatrix} \begin{bmatrix} x_1 \\ x_2 \end{bmatrix} = \begin{bmatrix} c_1 \\ c_2 \end{bmatrix} \qquad [1]$$

We can solve the equations by obtaining the inverse of the coefficients' matrix. In section 3.6.1 this was obtained as (equation [15], chapter 3)

$$\text{inverse of} \begin{bmatrix} a_{11} & a_{12} \\ a_{21} & a_{22} \end{bmatrix} = \frac{1}{a_{11}a_{22} - a_{12}a_{21}} \begin{bmatrix} a_{22} & -a_{12} \\ -a_{21} & a_{11} \end{bmatrix} \qquad [2]$$

The expression $a_{11}a_{22} - a_{12}a_{21}$ is known as the *determinant* of the matrix. This is represented as

$$\det \mathbf{A} = \begin{vmatrix} a_{11} & a_{12} \\ a_{21} & a_{22} \end{vmatrix} = a_{11}a_{22} - a_{12}a_{21} \qquad [3]$$

The determinant of the matrix is just written as the entries of the matrix in the same array as in the matrix but between | | brackets rather than the [ ] used for a matrix. The determinant, unlike a matrix which is an array of numbers, has a value, i.e. the number resulting from $a_{11}a_{22} - a_{12}a_{21}$. It is just another way of writing the expression $a_{11}a_{22} - a_{12}a_{21}$.

A useful way of remembering this relationship for the 2×2 matrix is by drawing diagonal lines through the matrix entries. Numbers on the same diagonal are multiplied together, with the diagonal from top left to bottom right giving a positive number and that from bottom left to top right a negative number.

$$
\begin{array}{cc}
+ & \\
\searrow & \nearrow \\
a_{11} & a_{12} \\
a_{21} & a_{22} \\
\nearrow & \searrow \\
- &
\end{array}
$$

The solution of the two simultaneous equations, i.e. the solution of the matrix equation [1], is thus given by

$$
\begin{bmatrix} x_1 \\ x_2 \end{bmatrix} = \frac{1}{\det \mathbf{A}} \begin{bmatrix} a_{22} & -a_{12} \\ -a_{21} & a_{11} \end{bmatrix} \begin{bmatrix} c_1 \\ c_2 \end{bmatrix}
$$

as

$$
x_1 = \frac{a_{22}c_1 - a_{12}c_2}{\det \mathbf{A}}
$$

$$
x_2 = \frac{a_{11}c_2 - a_{21}c_1}{\det \mathbf{A}}
$$

[4]

The above dealt with just a simple case of two simultaneous equations and so the determinant of a 2 × 2 matrix. We can, however, use determinants when we have far more simultaneous equations.

This chapter is about determinants and their properties, with chapter 5 being about their use in the determining of solutions to sets of simultaneous equations. Examples of their use in solving the simultaneous equations occurring with the analysis of electrical circuits are given in chapter 6.

Example

Determine the value of the determinant of the following matrix:

$$
\mathbf{A} = \begin{bmatrix} 4 & -1 \\ 2 & 3 \end{bmatrix}
$$

Using the relationship developed above, we can write the determinant of the matrix in the form

$$
\begin{array}{cc}
+ & \\
\searrow & \nearrow \\
4 & -1 \\
2 & 3 \\
\nearrow & \searrow \\
- &
\end{array}
$$

and hence we have

$$\det \mathbf{A} = (4 \times 3) - (2 \times -1) = 14$$

Review problems

1 Determine the values of the determinants of the following matrices:

(a) $\begin{bmatrix} 1 & 2 \\ 3 & 4 \end{bmatrix}$, (b) $\begin{bmatrix} -1 & 2 \\ 3 & 4 \end{bmatrix}$, (c) $\begin{bmatrix} 2 & -5 \\ 4 & 1 \end{bmatrix}$, (d) $\begin{bmatrix} 0 & 5 \\ 1 & 2 \end{bmatrix}$

4.1.1 Determinant of a 3 × 3 matrix

The concept of a determinant arises from the solution of simultaneous equations, as outlined in the previous section. However, we can define a determinant, in general terms, as an array of numbers that can be evaluated by particular rules. However, it must be said that the rules are determined by the relationships involved in the solution of simultaneous equations.

Consider the 3 × 3 matrix

$$\mathbf{A} = \begin{bmatrix} a_{11} & a_{12} & a_{13} \\ a_{21} & a_{22} & a_{23} \\ a_{31} & a_{32} & a_{33} \end{bmatrix}$$

The determinant of this matrix is defined as being

$$\det \mathbf{A} = a_{11}a_{22}a_{33} + a_{12}a_{23}a_{31} + a_{13}a_{21}a_{32} - a_{31}a_{22}a_{13} - a_{32}a_{23}a_{11} - a_{33}a_{21}a_{12} \tag{5}$$

As with the 2 × 2 matrix, a useful way of remembering this is by drawing diagonal lines through the matrix. So that the diagonal lines each pass through three entries, the first two columns are repeated to give the array shown below. The lines passing downwards from left to right give positive products of the entries, the lines passing upwards from left to right give negative products.

$$
\begin{array}{cccccc}
+ & + & + & & & \\
\searrow & \searrow & \searrow & \nearrow & \nearrow & \nearrow \\
a_{11} & a_{12} & a_{13} & a_{11} & a_{12} \\
a_{21} & a_{22} & a_{23} & a_{21} & a_{22} \\
a_{31} & a_{32} & a_{33} & a_{31} & a_{32} \\
\nearrow & \nearrow & \nearrow & \searrow & \searrow & \searrow \\
- & - & - & & &
\end{array}
$$

This method of evaluating determinants by multiplication of the numbers along the diagonals is only suitable for 2×2 and 3×3 determinants. It does *not work* for higher order determinants. Section 3.2 considers a method which can be used for determinants with more than 3 rows and 3 columns.

Example

Determine the value of the determinant of the following matrix **A**:

$$\begin{bmatrix} -1 & 2 & 3 \\ 4 & 5 & 6 \\ 7 & -8 & 9 \end{bmatrix}$$

Writing the array of numbers out in the way shown above,

$$\begin{array}{cccccc} + & + & + & & & \\ \searrow & \searrow & \searrow & \nearrow & \nearrow & \nearrow \\ -1 & 2 & 3 & -1 & 2 & \\ 4 & 5 & 6 & 4 & 5 & \\ 7 & -8 & 9 & 7 & -8 & \\ \nearrow & \nearrow & \nearrow & \searrow & \searrow & \searrow \\ - & - & - & & & \end{array}$$

Then we have

$$\begin{aligned} \det \mathbf{A} = &(-1 \times 5 \times 9) + (2 \times 6 \times 7) + (3 \times 4 \times -8) \\ &- (7 \times 5 \times 3) - (-8 \times 6 \times -1) - (9 \times 4 \times 2) \end{aligned}$$

$$= -45 + 84 - 96 - 105 - 48 - 72 = -282$$

Review problems

2 Determine the values of determinants of the following matrices:

(a) $\begin{bmatrix} 1 & 2 & 3 \\ -4 & 5 & 6 \\ 7 & -8 & 9 \end{bmatrix}$, (b) $\begin{bmatrix} 1 & 0 & 2 \\ -3 & 2 & 4 \\ 1 & 2 & 1 \end{bmatrix}$, (c) $\begin{bmatrix} 2 & 3 & 5 \\ 3 & 4 & 2 \\ 4 & -1 & 2 \end{bmatrix}$,

(d) $\begin{bmatrix} -1 & 3 & 5 \\ 2 & 0 & 1 \\ 1 & 1 & 0 \end{bmatrix}$, (e) $\begin{vmatrix} 3 & 0 & 0 \\ 1 & 5 & 7 \\ 4 & -2 & 3 \end{vmatrix}$

4.2 Minors and cofactors

Consider the 3×3 matrix, discussed in the previous section, of

$$\mathbf{A} = \begin{bmatrix} a_{11} & a_{12} & a_{13} \\ a_{21} & a_{22} & a_{23} \\ a_{31} & a_{32} & a_{33} \end{bmatrix}$$

and its determinant, given by equation [5],

$$\det \mathbf{A} = a_{11}a_{22}a_{33} + a_{12}a_{23}a_{31} + a_{13}a_{21}a_{32} \\ - a_{31}a_{22}a_{13} - a_{32}a_{23}a_{11} - a_{33}a_{21}a_{12} \tag{5}$$

We can rearrange these terms in a number of different ways, depending on which row or column entries we take outside the brackets. One way, in which we take outside brackets the entries of the first row, gives

$$\det \mathbf{A} = a_{11}(a_{22}a_{33} - a_{23}a_{32}) - a_{12}(a_{21}a_{33} - a_{23}a_{31}) \\ + a_{13}(a_{21}a_{32} - a_{22}a_{31}) \tag{6}$$

The quantities in the brackets are 2×2 determinants. Thus we can write equation [6] as

$$\det \mathbf{A} = a_{11}\begin{vmatrix} a_{22} & a_{23} \\ a_{32} & a_{33} \end{vmatrix} - a_{12}\begin{vmatrix} a_{21} & a_{23} \\ a_{31} & a_{33} \end{vmatrix}$$

$$+ a_{13}\begin{vmatrix} a_{21} & a_{22} \\ a_{31} & a_{32} \end{vmatrix} \tag{7}$$

Equation [6], and hence [7], involves multiplying each of the entries in the first row by 2×2 determinants. This has the entries in the first row multiplied by the determinants obtained by deleting the first row and the column containing the particular entry, i.e.

$$\det \mathbf{A} = a_{11}\begin{vmatrix} & & \\ a_{22} & a_{23} \\ a_{32} & a_{33} \end{vmatrix} - a_{12}\begin{vmatrix} a_{21} & a_{23} \\ a_{31} & a_{33} \end{vmatrix}$$

$$+ a_{13}\begin{vmatrix} a_{21} & a_{22} \\ a_{31} & a_{32} \end{vmatrix}$$

Each of the 2×2 determinants is called a *minor* in the determinant of $\mathbf{A}$ of the entry by which it is multiplied. Thus, for example,

$$\text{minor of } a_{11} \text{ in} \begin{vmatrix} a_{11} & a_{12} & a_{13} \\ a_{21} & a_{22} & a_{23} \\ a_{31} & a_{32} & a_{33} \end{vmatrix} = \begin{vmatrix} a_{22} & a_{23} \\ a_{32} & a_{33} \end{vmatrix}$$

Similar expressions for the 3×3 determinant can be obtained in which the minors are taken of the entries in any of the rows or columns. All that has happened is that equation [5] is rearranged in a different way. Equation [6] was one way, the following is another way, involving each of the entries in the first column being multiplied by 2×2 determinants.

$$\det \mathbf{A} = a_{11}(a_{22}a_{33} - a_{32}a_{23}) - a_{21}(a_{12}a_{33} - a_{32}a_{13}) \\ + a_{31}(a_{12}a_{23} - a_{22}a_{13}) \tag{8}$$

$$= a_{11} \begin{vmatrix} a_{22} & a_{23} \\ a_{32} & a_{33} \end{vmatrix} - a_{21} \begin{vmatrix} a_{12} & a_{13} \\ a_{32} & a_{33} \end{vmatrix}$$

$$+ a_{31} \begin{vmatrix} a_{12} & a_{13} \\ a_{22} & a_{23} \end{vmatrix}$$

Each minor of an entry has a sign associated with it, the sign depending on the position of the entry. The signs for a 3×3 determinant are given by the following pattern:

$$\begin{vmatrix} + & - & + \\ - & + & - \\ + & - & + \end{vmatrix} \tag{9}$$

The minor with its sign is termed a *cofactor*. Thus

$$\det \mathbf{A} = \text{sum of the entries of a column or row} \\ \text{multiplied by their cofactors} \tag{10}$$

This process of expressing a determinant as such a sum is termed the cofactor *expansion* of a row or column.

Example

Evaluate the following determinant using expansion by (a) the first row, (b) the first column:

$$\det \mathbf{A} = \begin{vmatrix} 2 & 3 & 4 \\ 1 & 2 & -3 \\ 1 & 1 & 5 \end{vmatrix}$$

(a) We can write this, using the signs given by [9], as

$$\det A = 2 \begin{vmatrix} 2 & -3 \\ 1 & 5 \end{vmatrix} - 3 \begin{vmatrix} 1 & -3 \\ 1 & 5 \end{vmatrix} + 4 \begin{vmatrix} 1 & 2 \\ 1 & 1 \end{vmatrix}$$

$$= 2(10 + 3) - 3(5 + 3) + 4(1 - 2) = -2$$

(b) We can write thus, using the signs given by [9], as

$$\det A = 2 \begin{vmatrix} 2 & -3 \\ 1 & 5 \end{vmatrix} - 1 \begin{vmatrix} 3 & 4 \\ 1 & 5 \end{vmatrix} + 1 \begin{vmatrix} 3 & 4 \\ 2 & -3 \end{vmatrix}$$

$$= 2(10 + 3) - 1(15 - 4) + 1(-9 - 8) = -2$$

Review problems

3 Determine the values of the cofactors of the entries in (a) the first row, (b) the second row, (c) the first column, of the following determinant:

$$\begin{vmatrix} 1 & 3 & 7 \\ 4 & -2 & 1 \\ 2 & 0 & 4 \end{vmatrix}$$

4 Evaluate the following 3×3 determinants using expansion by a row or column:

(a) $\begin{vmatrix} 2 & 5 & 3 \\ 1 & 7 & 0 \\ 5 & 2 & 4 \end{vmatrix}$, (b) $\begin{vmatrix} 1 & -2 & 3 \\ 4 & -5 & -1 \\ 0 & 1 & 4 \end{vmatrix}$, (c) $\begin{vmatrix} 3 & 1 & -1 \\ 5 & 2 & 0 \\ 6 & -2 & 1 \end{vmatrix}$,

(d) $\begin{vmatrix} 3 & 2 & -1 \\ -2 & -2 & 3 \\ 5 & 2 & 1 \end{vmatrix}$

4.2.1 $n \times n$ determinants

We can apply the cofactor method to determinants with 4×4 determinants and even higher numbers of rows and columns. As with the 3×3 determinants (equation [10]),

> det **A** = sum of the entries of a column or row
> multiplied by their cofactors

The signs for the cofactors are given by

$$
\begin{vmatrix}
+ & - & + & - & \cdots \\
- & + & - & + & \cdots \\
+ & - & + & - & \cdots \\
- & + & - & + & \cdots \\
\vdots & \vdots & \vdots & \vdots &
\end{vmatrix}
\qquad [11]
$$

Consider the 4×4 determinant

$$
\det \mathbf{A} = \begin{vmatrix}
a_{11} & a_{12} & a_{13} & a_{14} \\
a_{21} & a_{22} & a_{23} & a_{24} \\
a_{31} & a_{32} & a_{33} & a_{34} \\
a_{41} & a_{42} & a_{43} & a_{44}
\end{vmatrix}
$$

Expanding this determinant by the first row gives

$$
\det \mathbf{A} = a_{11}\begin{vmatrix}
a_{22} & a_{23} & a_{24} \\
a_{32} & a_{33} & a_{34} \\
a_{42} & a_{43} & a_{44}
\end{vmatrix}
$$

$$
+ a_{12}\begin{vmatrix}
a_{21} & a_{23} & a_{24} \\
a_{31} & a_{33} & a_{34} \\
a_{41} & a_{43} & a_{44}
\end{vmatrix}
$$

$$
+ a_{13}\begin{vmatrix}
a_{21} & a_{22} & a_{24} \\
a_{31} & a_{32} & a_{34} \\
a_{41} & a_{42} & a_{44}
\end{vmatrix}
$$

$$
+ a_{14}\begin{vmatrix}
a_{21} & a_{22} & a_{23} \\
a_{31} & a_{32} & a_{33} \\
a_{41} & a_{42} & a_{43}
\end{vmatrix}
\qquad [12]
$$

The 3×3 determinants can be evaluated by using the cofactor method or the multiplication along diagonals method. Hence the 4×4 determinant can be evaluated. The same method can be used for determinants with yet more rows and columns.

Example

Evaluate the following 4×4 determinant:

$$\begin{vmatrix} 1 & 3 & 0 & 0 \\ 3 & 1 & 0 & 0 \\ 0 & 0 & 1 & -4 \\ 0 & 0 & 7 & 9 \end{vmatrix}$$

Expanding this determinant by the first row gives

$$\det \mathbf{A} = 1 \begin{vmatrix} 1 & 0 & 0 \\ 0 & 1 & -4 \\ 0 & 7 & 9 \end{vmatrix} - 3 \begin{vmatrix} 3 & 0 & 0 \\ 0 & 1 & -4 \\ 0 & 7 & 9 \end{vmatrix}$$

$$+ 0 + 0$$

Using the diagonal rule to evaluate the 3×3 determinants,

$$\det \mathbf{A} = 1(9 + 28) - 3(27 + 84) = -296$$

We could have expanded the determinant by the first column and obtained

$$\det \mathbf{A} = 1 \begin{vmatrix} 1 & 0 & 0 \\ 0 & 1 & -4 \\ 0 & 7 & 9 \end{vmatrix} - 3 \begin{vmatrix} 3 & 0 & 0 \\ 0 & 1 & -4 \\ 0 & 7 & 9 \end{vmatrix}$$

$$= 1(9 + 28) - 3(27 + 84) = -296$$

Review problems

5   Evaluate the following determinants using expansion by either a row or a column:

(a) $\begin{vmatrix} 3 & 2 & 0 & 1 \\ 7 & 5 & 2 & 2 \\ -1 & 0 & 6 & 0 \\ 3 & 2 & 3 & 3 \end{vmatrix}$, (b) $\begin{vmatrix} 1 & 3 & 5 & 7 \\ 2 & 4 & 6 & 8 \\ 3 & 5 & 7 & 9 \\ 4 & 6 & 8 & 10 \end{vmatrix}$,

$$(c) \begin{vmatrix} 1 & 1 & 1 & 1 \\ 0 & 1 & 1 & 1 \\ 0 & 0 & 1 & 1 \\ 0 & 0 & 0 & 1 \end{vmatrix}, (d) \begin{vmatrix} 1 & a & a^2 & 0 \\ 0 & 1 & a & a^2 \\ a^2 & 0 & 1 & a \\ a & a^2 & 0 & 1 \end{vmatrix}$$

4.3 Properties of determinants

The following are properties of determinants which can be of use in simplifying them in order to make calculations easier.

4.3.1 Interchanging all rows with columns

Consider the following determinant:

$$\det \mathbf{A} = \begin{vmatrix} a_{11} & a_{12} \\ a_{21} & a_{22} \end{vmatrix} = a_{11}a_{22} - a_{21}a_{12}$$

If we now change all the rows to columns we have the determinant

$$\det \mathbf{B} = \begin{vmatrix} a_{11} & a_{21} \\ a_{12} & a_{22} \end{vmatrix} = a_{11}a_{22} - a_{21}a_{12}$$

The value of the determinant is unchanged, det $\mathbf{A}$ = det $\mathbf{B}$. A determinant is unaltered in value if the rows are changed to columns.

Example

Which of the following determinants are equal in value?

$$\det \mathbf{A} = \begin{vmatrix} 2 & 4 & 7 \\ 1 & 4 & 2 \\ 4 & 1 & 3 \end{vmatrix}$$

$$\det \mathbf{B} = \begin{vmatrix} 4 & 7 & 2 \\ 4 & 2 & 1 \\ 1 & 3 & 1 \end{vmatrix}$$

$$\det \mathbf{C} = \begin{vmatrix} 2 & 1 & 4 \\ 4 & 4 & 1 \\ 7 & 2 & 3 \end{vmatrix}$$

Det $\mathbf{A}$ and det $\mathbf{C}$ are equal in value since the rows in $\mathbf{A}$ have just been transformed into columns in $\mathbf{C}$. Det $\mathbf{B}$ is a rearrangement of rows and this does not give a value the same as that of $\mathbf{A}$.

Review problems

6   Which of the following determinants are equal in value?

$$\det \mathbf{A} = \begin{vmatrix} 1 & 4 \\ -1 & 2 \end{vmatrix}, \quad \det \mathbf{B} = \begin{vmatrix} 1 & -1 \\ 4 & 2 \end{vmatrix}, \quad \det \mathbf{C} = \begin{vmatrix} 4 & 1 \\ 2 & -1 \end{vmatrix}$$

4.3.2 Interchanging rows or columns

Consider the following determinant:

$$\det \mathbf{A} = \begin{vmatrix} a_{11} & a_{12} \\ a_{21} & a_{22} \end{vmatrix} = a_{11}a_{22} - a_{21}a_{12}$$

If we interchange rows 1 and 2 and have a new determinant, then

$$\det \mathbf{B} = \begin{vmatrix} a_{21} & a_{22} \\ a_{11} & a_{12} \end{vmatrix} = a_{21}a_{12} - a_{11}a_{22}$$

Thus we have $\det \mathbf{B} = -\det \mathbf{A}$. Interchanging the rows has just changed the sign of the value of the determinant.

In general, if a determinant $\mathbf{B}$ is obtained from $\mathbf{A}$ by interchanging two rows, or two columns, of $\mathbf{A}$, then

$$\det \mathbf{B} = -\det \mathbf{A} \qquad\qquad [13]$$

Note that if there are two interchanges the determinant will change sign twice and hence remain at its original value.

Example

If the determinant

$$\begin{vmatrix} 1 & 2 & 0 \\ 0 & 3 & 1 \\ 2 & 1 & 1 \end{vmatrix}$$

has the value of 6, determine the values of the following determinants:

$$\text{(a)} \begin{vmatrix} 1 & 2 & 0 \\ 2 & 1 & 1 \\ 0 & 3 & 1 \end{vmatrix}, \quad \text{(b)} \begin{vmatrix} 0 & 2 & 1 \\ 1 & 3 & 0 \\ 1 & 1 & 2 \end{vmatrix}$$

(a) This determinant is just the original determinant with the second and third rows interchanged. Thus the value is −6.

(b) This determinant is just the original determinant with the first and third columns interchanged. Thus the value is −6.

Review problems

7 If the determinant

$$\begin{vmatrix} 0 & 1 & 1 \\ 2 & 2 & 1 \\ 3 & 2 & 1 \end{vmatrix}$$

has the value −1, determine the values of the following determinants:

$$\text{(a)} \begin{vmatrix} 0 & 1 & 1 \\ 3 & 2 & 1 \\ 2 & 2 & 1 \end{vmatrix}, \text{(b)} \begin{vmatrix} 1 & 0 & 1 \\ 2 & 2 & 1 \\ 2 & 3 & 1 \end{vmatrix}, \text{(c)} \begin{vmatrix} 3 & 2 & 1 \\ 0 & 1 & 1 \\ 2 & 2 & 1 \end{vmatrix}$$

4.3.3 Equal rows or columns

Consider the following determinant, in which two rows are the same:

$$\det \mathbf{A} = \begin{vmatrix} a_1 & b_1 & c_1 \\ a_1 & b_1 & c_1 \\ a_2 & b_2 & c_2 \end{vmatrix}$$

If we interchange the equal rows, i.e. rows 1 and 2, then we have

$$\det \mathbf{B} = \begin{vmatrix} a_1 & b_1 & c_1 \\ a_1 & b_1 & c_1 \\ a_2 & b_2 & c_2 \end{vmatrix}$$

But we must have, equation [13], $\det \mathbf{B} = -\det \mathbf{A}$. This can only be the case, since **B** and **A** are identical, if $\det \mathbf{A} = \det \mathbf{B} = 0$. The same result occurs when two columns are equal.

Example

What must be the value of x for the following determinant to have a zero value?

$$\det \mathbf{A} = \begin{vmatrix} x & 2 & x+3 \\ 2 & 2x & 4 \\ 3 & 2 & 6 \end{vmatrix} = 0$$

There must be two rows or two columns identical. If $x = 3$ the first and third row are identical.

4.3.4 Multiplication of a row, or column, by a constant

Consider the following determinant and the result of multiplying the first row by a constant, i.e. 3 in this case:

$$\det \mathbf{A} = \begin{vmatrix} a_1 & a_2 \\ b_1 & b_2 \end{vmatrix} = a_1 b_2 - b_1 a_2$$

$$\det \mathbf{B} = \begin{vmatrix} 3a_1 & 3a_2 \\ b_1 & b_2 \end{vmatrix} = 3a_1 b_2 - 3b_1 a_2$$

Thus

$$\det \mathbf{B} = 3 \det \mathbf{A}$$

If the entries in a row, or a column, of a determinant $\mathbf{A}$ are each multiplied by the same factor k then the result is the product of that factor and the original determinant.

$$\det \mathbf{B} = k \det \mathbf{A} \tag{14}$$

Example

If the determinant

$$\begin{vmatrix} 2 & 3 \\ 1 & 4 \end{vmatrix} = 5$$

determine the value of the determinant

$$\begin{vmatrix} 6 & 3 \\ 3 & 4 \end{vmatrix}$$

The entries in the first column are three times those in the original determinant, the other columns being identical. Hence the determinant has the value $3 \times 5 = 15$.

Example

Simplify the following determinant by factoring it.

$$\begin{vmatrix} 9 & 15 \\ -2 & 25 \end{vmatrix}$$

Extracting a factor of 3 from the first row gives

$$\begin{vmatrix} 9 & 15 \\ -2 & 25 \end{vmatrix} = 3 \begin{vmatrix} 3 & 5 \\ -2 & 25 \end{vmatrix}$$

Extracting a factor of 5 from the second column gives

$$\begin{vmatrix} 9 & 15 \\ -2 & 25 \end{vmatrix} = 3 \times 5 \begin{vmatrix} 3 & 1 \\ -2 & 5 \end{vmatrix} = 3 \times 5 \times (15 + 2) = 255$$

Review problems

8 Determine the value of x which will make these two determinants equal.

$$\begin{vmatrix} 1 & 4 \\ 2 & 6 \end{vmatrix} = 2 \begin{vmatrix} 1 & x \\ 2 & 3 \end{vmatrix}$$

9 Simplify the following determinant by extracting factors:

$$\begin{vmatrix} 4 & 24 \\ 2 & 3 \end{vmatrix}$$

4.3.5 A row or column is a multiple of another

Consider a determinant where one row is a multiple of another row. For example, the following determinant has row 2 as twice row 1,

$$\begin{vmatrix} 1 & 2 & 3 \\ 2 & 4 & 6 \\ 1 & 3 & 1 \end{vmatrix}$$

We can extract the factor 2 (see section 4.3.4) and thus write

$$2 \begin{vmatrix} 1 & 2 & 3 \\ 1 & 2 & 3 \\ 1 & 3 & 1 \end{vmatrix}$$

Because rows 1 and 2 are the same, the determinant has a value of zero (see section 4.3.3). Thus, when a row, or column, is a multiple of another then the value of the determinant is zero.

4.3.6 All the entries in a row or column are zero

Consider the following determinant in which all the entries in the second row are 0:

$$\begin{vmatrix} 1 & 2 & 3 \\ 0 & 0 & 0 \\ 4 & 5 & 6 \end{vmatrix}$$

Since there is a zero on each diagonal then when it is multiplied out the result must be 0. The same situation occurs if all the entries in a column are zero.

If all the entries of a row or a column of a determinant are zero then the value of that determinant is zero.

4.3.7 Addition of rows or columns

Consider the following determinant:

$$\begin{vmatrix} a_1 & a_2 & a_3 \\ b_1 & b_2 & b_3 \\ c_1 & c_2 & c_3 \end{vmatrix}$$

Evaluation of this gives

$$a_1 b_2 c_3 + a_2 b_3 c_1 + a_3 b_1 c_2 - c_1 b_2 a_3 - c_2 b_3 a_1 - c_3 b_1 a_1$$

We now rewrite the determinant with the second row having k times the third row added to it.

$$\begin{vmatrix} a_1 & a_2 & a_3 \\ b_1 + kc_1 & b_2 + kc_2 & b_3 + kc_3 \\ c_1 & c_2 & c_3 \end{vmatrix}$$

Evaluation of this gives

$$a_1 b_2 c_3 + ka_1 c_2 c_3 + a_2 b_3 c_1 + ka_2 c_3 c_1 + a_3 b_1 c_2 + ka_3 c_1 c_2$$
$$- c_1 b_2 a_3 - kc_1 c_2 a_3 - c_2 b_3 a_1 - kc_2 c_3 a_1 - c_3 b_1 a_2 - kc_3 c_1 a_2$$

Grouping the terms gives, for those not including k the original determinant, and for those terms involving k

$$k[c_1(a_2 c_3 - c_3 a_2) + c_2(a_1 c_3 - c_3 a_1) + c_3(a_1 c_2 - c_2 a_1)]$$

These terms give a zero sum. Hence, adding a multiple of one row or column to another does not affect the value of the determinant.

Consider the following numerical example:

$$\begin{vmatrix} 2 & -1 & 2 \\ -2 & 3 & 4 \\ -1 & 2 & 2 \end{vmatrix}$$

We can modify the determinant by adding the twice the third row to the first row, with the result

$$\begin{vmatrix} 0 & 3 & 6 \\ -2 & 3 & 4 \\ -1 & 2 & 2 \end{vmatrix}$$

We could further modify the determinant by adding (-2) times the third row to the second row, with the result

$$\begin{vmatrix} 0 & 3 & 6 \\ 0 & -1 & 0 \\ -1 & 2 & 2 \end{vmatrix}$$

The result is a simplified determinant which is simpler to evaluate. The introduction of zeros considerably reduces the arithmetic involved when multiplying out the answer.

Example

Evaluate, after simplification, the following determinant:

$$\begin{vmatrix} 1 & 1 & 8 \\ 0 & 4 & 10 \\ -2 & 1 & 6 \end{vmatrix}$$

Taking a factor 2 out of the third column gives

$$2 \begin{vmatrix} 1 & 1 & 4 \\ 0 & 4 & 5 \\ -2 & 1 & 3 \end{vmatrix}$$

Adding (-1) times the first column to the second column gives

$$2 \begin{vmatrix} 1 & 0 & 4 \\ 0 & 4 & 5 \\ -2 & 3 & 3 \end{vmatrix}$$

Adding (–4) times the first column to the third column gives

$$2\begin{vmatrix} 1 & 0 & 0 \\ 0 & 4 & 5 \\ -2 & 3 & 11 \end{vmatrix}$$

Hence the determinant has the value

$$2\,(1 \times 4 \times 11 - 3 \times 5 \times 1) = 58$$

Example

Determine the value of x which satisfies the following equation:

$$\begin{vmatrix} x & 1 & 1 \\ 1 & x & 1 \\ 1 & 1 & x \end{vmatrix} = 0$$

Subtracting the second row from the first row gives

$$\begin{vmatrix} x-1 & 1-x & 0 \\ 1 & x & 1 \\ 1 & 1 & x \end{vmatrix} = 0$$

Adding the first column to the second column gives

$$\begin{vmatrix} x-1 & 0 & 0 \\ 1 & x+1 & 1 \\ 1 & 2 & x \end{vmatrix} = 0$$

If we let $x = 1$ then the first row becomes all zeros, hence the determinant will have a zero value. Thus a solution is $x = 1$. Multiplying out the determinant along the diagonals gives

$$(x - 1)(x + 1)x - 2(x - 1) = 0$$

$$x^3 - 3x + 2 = 0$$

We know that one of the factors must be $(x - 1)$ and so we have

$$(x^2 + x - 2)(x - 1) = 0$$

$$(x + 2)(x - 1)(x - 1) = 0$$

Thus $x = 1$ or -2.

Review problems

10 If we have

$$\begin{vmatrix} a_1 & a_2 & a_3 \\ b_1 & b_2 & b_3 \\ c_1 & c_2 & c_3 \end{vmatrix} = 12$$

determine the values of the following determinants:

(a) $\begin{vmatrix} a_1 & a_2 & a_3 \\ b_1 + a_1 & b_2 + a_2 & b_3 + a_3 \\ c_1 & c_2 & c_3 \end{vmatrix}$,

(b) $\begin{vmatrix} a_1 + 2a_2 & a_2 & a_3 \\ b_1 + 2b_2 & b_2 & b_3 \\ c_1 + 2c_2 & c_2 & c_3 \end{vmatrix}$,

(c) $\begin{vmatrix} 2a_1 & 2a_2 & 2a_3 \\ b_1 & b_2 & b_3 \\ c_1 - 2a_1 & c_2 - 2a_2 & c_3 - 2a_3 \end{vmatrix}$,

(d) $\begin{vmatrix} b_1 - b_2 & b_2 & b_3 \\ a_1 - a_2 & a_2 & a_3 \\ c_1 - c_2 & c_2 & c_3 \end{vmatrix}$

11 Evaluate, after simplification, the following determinants:

(a) $\begin{vmatrix} 4 & 2 & 3 \\ 5 & 7 & 5 \\ 4 & 7 & 6 \end{vmatrix}$, (b) $\begin{vmatrix} 9 & 15 & 21 \\ 11 & 9 & 13 \\ 15 & 17 & 19 \end{vmatrix}$, (c) $\begin{vmatrix} -5 & 5 & 4 \\ -4 & 5 & 5 \\ 5 & -1 & 2 \end{vmatrix}$

12 Show that

$$\begin{vmatrix} b+c & c+a & a+b \\ m+n & n+l & l+m \\ q+r & r+p & p+q \end{vmatrix} = 2 \begin{vmatrix} a & b & c \\ l & m & n \\ p & q & r \end{vmatrix}$$

13 Show that

$$\begin{vmatrix} 1 & 1 & 1 \\ x & y & z \\ x^2 & y^2 & z^2 \end{vmatrix} = (y-x)(z-x)(z-y)$$

14 Determine x in the following equation:

$$\begin{vmatrix} x+1 & x+2 & 3 \\ 2 & x+3 & x+1 \\ x+3 & 1 & x+2 \end{vmatrix} = 0$$

Further problems

15 Determine, using the rules for diagonal multiplication, the values of the determinants of the following matrices:

(a) $\begin{bmatrix} 2 & 4 \\ 1 & -1 \end{bmatrix}$, (b) $\begin{bmatrix} -1 & 5 \\ 2 & 3 \end{bmatrix}$, (c) $\begin{bmatrix} -2 & -1 \\ 4 & -1 \end{bmatrix}$,

(d) $\begin{bmatrix} \cos\theta & -\sin\theta \\ \sin\theta & \cos\theta \end{bmatrix}$, (e) $\begin{bmatrix} 1 & 2 & 0 \\ -1 & 1 & 2 \\ 2 & 1 & -1 \end{bmatrix}$,

(f) $\begin{bmatrix} -1 & 0 & 2 \\ 2 & 0 & 1 \\ 1 & -1 & 2 \end{bmatrix}$, (g) $\begin{bmatrix} 2 & 4 & 1 \\ 1 & 0 & -2 \\ -1 & 2 & 1 \end{bmatrix}$, (h) $\begin{bmatrix} 3 & 4 & 2 \\ 2 & -3 & 1 \\ 1 & 5 & 1 \end{bmatrix}$

16 Evaluate the following determinants using expansion by the first row:

(a) $\begin{vmatrix} 1 & 2 & 3 \\ 2 & 3 & 0 \\ 3 & 0 & 0 \end{vmatrix}$, (b) $\begin{vmatrix} 1 & 2 & 3 \\ 0 & 1 & 2 \\ 1 & 0 & 3 \end{vmatrix}$, (c) $\begin{vmatrix} 1 & 2 & 1 \\ 0 & 1 & 0 \\ 1 & 1 & 0 \end{vmatrix}$,

(d) $\begin{vmatrix} 1 & 2 & -3 \\ 2 & -3 & 4 \\ 3 & 4 & 5 \end{vmatrix}$, (e) $\begin{vmatrix} 1 & 0 & 0 & 1 \\ 1 & 2 & 1 & 1 \\ 0 & 1 & 2 & 1 \\ 0 & 0 & 1 & 2 \end{vmatrix}$,

(f) $\begin{vmatrix} 2 & 0 & 1 & 2 \\ 1 & 1 & 0 & 2 \\ 2 & -1 & 3 & 1 \\ 3 & -1 & 4 & 3 \end{vmatrix}$, (g) $\begin{vmatrix} 2 & 2 & 3 & 1 \\ 0 & 1 & 4 & 2 \\ 0 & 0 & 1 & 5 \\ 1 & 3 & 3 & 0 \end{vmatrix}$,

(h) $\begin{vmatrix} 2 & 1 & 0 & 0 & 0 \\ 1 & 2 & 1 & 0 & 0 \\ 0 & 1 & 2 & 1 & 0 \\ 0 & 0 & 1 & 2 & 1 \\ 0 & 0 & 0 & 1 & 2 \end{vmatrix}$

17 Evaluate, after simplification, the following determinants:

(a) $\begin{vmatrix} 3 & 5 & 7 \\ 11 & 9 & 13 \\ 15 & 17 & 19 \end{vmatrix}$, (b) $\begin{vmatrix} 10 & 12 & 13 \\ 12 & 13 & 15 \\ 14 & 15 & 16 \end{vmatrix}$,

(c) $\begin{vmatrix} 22 & 25 & 28 \\ 26 & 28 & 31 \\ 24 & 27 & 29 \end{vmatrix}$, (d) $\begin{vmatrix} 2 & 2 & 5 & 9 \\ 3 & 2 & 9 & 5 \\ 4 & -3 & 7 & 7 \\ 2 & 9 & 7 & 15 \end{vmatrix}$,

(e) $\begin{vmatrix} 3 & 5 & 3 & 2 \\ 4 & 3 & -1 & 4 \\ 8 & 7 & 3 & 6 \\ 11 & 10 & -4 & 8 \end{vmatrix}$

18 Determine x in the following determinant:

$$\begin{vmatrix} x-3 & x+2 & x-1 \\ x+2 & x-4 & x \\ x-1 & x+4 & x-5 \end{vmatrix} = 0$$

19 Prove that

$$\begin{vmatrix} a^2 & a & bc \\ b^2 & b & ca \\ c^2 & c & ab \end{vmatrix} = \begin{vmatrix} a^3 & a^2 & 1 \\ b^3 & b^2 & 1 \\ c^3 & c^2 & 1 \end{vmatrix}$$

20 Given that

$$\begin{vmatrix} a_1 & a_2 & a_3 \\ b_1 & b_2 & b_3 \\ c_1 & c_2 & c_3 \end{vmatrix} = 10$$

determine the values of the following determinants:

(a) $\begin{vmatrix} a_2 & a_1 & a_3 \\ b_2 & b_1 & b_3 \\ c_2 & c_1 & c_3 \end{vmatrix}$, (b) $\begin{vmatrix} a_1 & a_2 & a_3 \\ 2b_1 & 2b_2 & 2b_3 \\ c_1 & c_2 & c_3 \end{vmatrix}$,

(c) $\begin{vmatrix} a_1 & 3a_2 & a_3 \\ b_1 & 3b_2 & b_3 \\ c_1 & 3c_2 & c_3 \end{vmatrix}$, (d) $\begin{vmatrix} a_1 & a_2 & a_3 \\ c_1 & c_2 & c_3 \\ b_1 & b_2 & b_3 \end{vmatrix}$,

$$\text{(e)} \begin{vmatrix} a_2 & a_3 & a_1 \\ b_2 & b_3 & b_1 \\ c_2 & c_3 & c_1 \end{vmatrix}, \text{(f)} \begin{vmatrix} a_1 - b_1 & a_2 - b_2 & a_3 - b_3 \\ b_1 - c_1 & b_2 - c_2 & b_3 - c_3 \\ c_1 & c_2 & c_3 \end{vmatrix},$$

$$\text{(g)} \begin{vmatrix} a_1 - 2b_1 & a_2 - 2b_2 & a_3 - 2b_3 \\ b_1 & b_2 & b_3 \\ c_1 & c_2 & c_3 \end{vmatrix}$$

5 Determinants and inverses

5.1 Inverse matrices

This chapter is concerned with determinants as a means of determining the inverse of matrices, and hence obtaining the solution of simultaneous linear equations. Arising from this is the development of a formula, Cramer's rule, for the solution of simultaneous linear equations.

Inversion of a matrix was introduced in section 3.6 in relation to the solution of simultaneous linear equations. Consider two such equations with two variables, x_1 and x_2,

$$a_{11}x_1 + a_{12}x_2 = c_1$$

$$a_{21}x_1 + a_{22}x_2 = c_2$$

The a and c terms are constants. We can represent these equations in terms of matrices as (chapter 3, equation [11])

$$\begin{bmatrix} a_{11} & a_{12} \\ a_{21} & a_{22} \end{bmatrix} \begin{bmatrix} x_1 \\ x_2 \end{bmatrix} = \begin{bmatrix} c_1 \\ c_2 \end{bmatrix}$$

or

$$\mathbf{Ax} = \mathbf{c} \qquad [1]$$

We can define the *inverse of a matrix* $\mathbf{A}^{-1}$ as that which when multiplied by $\mathbf{A}$ gives the unit matrix $\mathbf{I}$ (chapter 3, equation [13]), i.e.

$$\mathbf{AA}^{-1} = \mathbf{I} \qquad [2]$$

Thus, multiplying both sides of equation [1] by $\mathbf{A}^{-1}$ gives

$$\mathbf{A}^{-1}\mathbf{Ax} = \mathbf{A}^{-1}\mathbf{c}$$

Since $\mathbf{A}^{-1}\mathbf{A} = \mathbf{A}\mathbf{A}^{-1} = \mathbf{I}$ (chapter 3, equation [10]) and the value of the unit matrix is 1, then we have

$$\mathbf{x} = \mathbf{A}^{-1}\mathbf{c} \qquad [3]$$

This equation represents the solutions of the simultaneous equations. Thus, by determining the inverse of a matrix we are able to solve the simultaneous equations.

In section 3.6.2 row operations were used to determine the inverse of a matrix. In this chapter another method is considered.

5.2 Cofactor and adjoint matrices

Consider three simultaneous equations in three variables,

$$a_{11}x_1 + a_{12}x_2 + a_{13}x_3 = c_1$$

$$a_{21}x_1 + a_{22}x_2 + a_{23}x_3 = c_2$$

$$a_{31}x_1 + a_{32}x_2 + a_{33}x_3 = c_3$$

The coefficient matrix is

$$\mathbf{A} = \begin{bmatrix} a_{11} & a_{12} & a_{13} \\ a_{21} & a_{22} & a_{23} \\ a_{31} & a_{32} & a_{33} \end{bmatrix} \qquad [4]$$

The determinant of this matrix is given by (see section 4.1)

$$\det \mathbf{A} = \begin{vmatrix} a_{11} & a_{12} & a_{13} \\ a_{21} & a_{22} & a_{23} \\ a_{31} & a_{32} & a_{33} \end{vmatrix} \qquad [5]$$

This determinant can be written in terms of *cofactors* (see section 4.2). The cofactor of a_{11} is

$$A_{11} = \begin{vmatrix} a_{22} & a_{23} \\ a_{32} & a_{33} \end{vmatrix}$$

The convention here is that of using italic capital letters for cofactors and giving them the subscript of the entry to which they refer. The cofactor of a_{12} is

$$A_{12} = - \begin{vmatrix} a_{21} & a_{23} \\ a_{31} & a_{33} \end{vmatrix}$$

See chapter 4 and the patterns given by [9] and [11] for the signs given to cofactors. The value of the determinant of **A** is given by expansion of the determinant about one row or one column (see chapter 4, equation [10])

det **A** = sum of the entries of a column or row
multiplied by their cofactors   [6]

Thus

$$\det \mathbf{A} = a_{11}A_{11} + a_{12}A_{12} + a_{13}A_{13}$$

Note that negative signs are incorporated with the cofactors and so we are concerned with just the sum of the terms.

Since, with the three simultaneous equations, there are nine coefficients there will be nine cofactors. We can write the *cofactor matrix* of the matrix **A** as

$$\text{cofactor } \mathbf{A} = \begin{bmatrix} A_{11} & A_{12} & A_{13} \\ A_{21} & A_{22} & A_{23} \\ A_{31} & A_{32} & A_{33} \end{bmatrix} \qquad [7]$$

We have another matrix, called the *adjoint matrix* or *adjugate matrix*, which we define as being the transpose of the cofactor matrix (see section 3.2.3 for a discussion of the transpose of a matrix). The transpose of a matrix is obtained by interchanging the rows and columns. Thus

$$\text{adj } \mathbf{A} = \begin{bmatrix} A_{11} & A_{21} & A_{31} \\ A_{12} & A_{22} & A_{32} \\ A_{13} & A_{23} & A_{33} \end{bmatrix} \qquad [8]$$

It is this adjoint matrix which is required for the determination of the inverse of a matrix.

Example

Determine the cofactor matrix and the adjoint matrix of the following matrix:

$$\begin{bmatrix} 1 & 2 & 4 \\ 0 & -1 & -2 \\ 1 & 3 & 2 \end{bmatrix}$$

(a) The cofactors of the matrix are given by

$$A_{11} = + \begin{vmatrix} -1 & -2 \\ 3 & 2 \end{vmatrix} = -2 + 6 = 4$$

$$A_{12} = - \begin{vmatrix} 0 & -2 \\ 1 & 2 \end{vmatrix} = -2$$

$$A_{13} = + \begin{vmatrix} 0 & -1 \\ 1 & 3 \end{vmatrix} = 1$$

$$A_{21} = - \begin{vmatrix} 2 & 4 \\ 3 & 2 \end{vmatrix} = -4 + 12 = 8$$

$$A_{22} = + \begin{vmatrix} 1 & 4 \\ 1 & 2 \end{vmatrix} = 2 - 4 = -2$$

$$A_{23} = - \begin{vmatrix} 1 & 2 \\ 1 & 3 \end{vmatrix} = -3 + 2 = -1$$

$$A_{31} = + \begin{vmatrix} 2 & 4 \\ -1 & -2 \end{vmatrix} = -4 + 4 = 0$$

$$A_{32} = - \begin{vmatrix} 1 & 4 \\ 0 & -2 \end{vmatrix} = 2$$

$$A_{33} = + \begin{vmatrix} 1 & 2 \\ 0 & -1 \end{vmatrix} = -1$$

Thus the cofactor matrix is

$$\begin{bmatrix} 4 & -2 & 1 \\ 8 & -2 & -1 \\ 0 & 2 & -1 \end{bmatrix}$$

(b) The adjoint matrix is obtained by transposing the cofactor matrix. Thus the adjoint matrix is

$$\begin{bmatrix} 4 & 8 & 0 \\ -2 & -2 & 2 \\ 1 & -1 & -1 \end{bmatrix}$$

Review problems

1   Determine the cofactor matrix and the adjoint matrix for each of the following matrices:

$$(a) \begin{bmatrix} 1 & 0 & 0 \\ 0 & 0 & 1 \\ 0 & 1 & 0 \end{bmatrix}, \quad (b) \begin{bmatrix} 2 & -1 & 3 \\ 0 & 4 & -2 \\ 1 & -3 & 5 \end{bmatrix}$$

5.2.1 The inverse matrix

Consider the product of a matrix and its adjoint matrix, i.e.

$$\mathbf{A}(\text{adj } \mathbf{A}) = \begin{bmatrix} a_{11} & a_{12} & a_{13} \\ a_{21} & a_{22} & a_{23} \\ a_{31} & a_{32} & a_{33} \end{bmatrix} \begin{bmatrix} A_{11} & A_{21} & A_{31} \\ A_{12} & A_{22} & A_{32} \\ A_{13} & A_{23} & A_{33} \end{bmatrix}$$

Suppose this product is a matrix $\mathbf{B}$, i.e. $\mathbf{A}(\text{adj } \mathbf{A}) = \mathbf{B}$, with

$$\mathbf{B} = \begin{bmatrix} b_{11} & b_{12} & b_{13} \\ b_{21} & b_{22} & b_{23} \\ b_{31} & b_{32} & b_{33} \end{bmatrix}$$

Then we must have

$$b_{11} = a_{11}A_{11} + a_{12}A_{12} + a_{13}A_{13}$$

$$b_{12} = a_{11}A_{21} + a_{12}A_{22} + a_{13}A_{23}$$

$$b_{13} = a_{11}A_{31} + a_{12}A_{32} + a_{13}A_{33}$$

$$b_{21} = a_{21}A_{11} + a_{22}A_{12} + a_{23}A_{13}$$

$$b_{22} = a_{21}A_{21} + a_{22}A_{22} + a_{23}A_{23}$$

$$b_{23} = a_{21}A_{31} + a_{22}A_{32} + a_{23}A_{33}$$

$$b_{31} = a_{31}A_{11} + a_{32}A_{12} + a_{33}A_{13}$$

$$b_{32} = a_{31}A_{21} + a_{32}A_{22} + a_{33}A_{23}$$

$$b_{33} = a_{31}A_{31} + a_{32}A_{32} + a_{33}A_{33}$$

b_{11}, b_{22} and b_{33} are the entries of A multiplied by the cofactors of the same entries. Thus

$$b_{11} = b_{22} = b_{33} = \det \mathbf{A}$$

With all the other entries, the 'a's and the cofactors come from different rows of A. Because of this, they are all zero. To illustrate this, consider b_{21}.

$$b_{21} = a_{21}A_{11} + a_{22}A_{12} + a_{23}A_{13}$$

$$\begin{aligned} = a_{21}(a_{22}a_{33} - a_{32}a_{23}) - a_{22}(a_{21}a_{33} - a_{31}a_{23}) \\ + a_{23}(a_{21}a_{32} - a_{31}a_{23}) \end{aligned}$$

$$= 0$$

Thus

$$\mathbf{A}(\text{adj } \mathbf{A}) = \begin{bmatrix} \det \mathbf{A} & 0 & 0 \\ 0 & \det \mathbf{A} & 0 \\ 0 & 0 & \det \mathbf{A} \end{bmatrix}$$

If we take the factor det A out of the matrix we have

$$\mathbf{A}(\text{adj } \mathbf{A}) = \det \mathbf{A} \begin{bmatrix} 1 & 0 & 0 \\ 0 & 1 & 0 \\ 0 & 0 & 1 \end{bmatrix}$$

Hence

$$\mathbf{A}(\text{adj } \mathbf{A}) = (\det \mathbf{A})\mathbf{I}$$

Dividing both sides of the equation by det A gives

$$\mathbf{I} = \mathbf{A}\,\frac{\text{adj } \mathbf{A}}{\det \mathbf{A}}$$

But $\mathbf{A}\mathbf{A}^{-1} = \mathbf{I}$, hence

$$\mathbf{A}^{-1} = \frac{\text{adj } \mathbf{A}}{\det \mathbf{A}} \qquad\qquad [9]$$

Note that if det A = 0 that there can be no inverse matrix.

Example

Determine the inverse of the matrix

$$\mathbf{A} = \begin{bmatrix} 1 & 2 & 4 \\ 0 & -1 & -2 \\ 1 & 3 & 2 \end{bmatrix}$$

This is the example considered in the previous section. The adjoint matrix was there found to be

$$\text{adj } \mathbf{A} = \begin{bmatrix} 4 & 8 & 0 \\ -2 & -2 & 2 \\ 1 & -1 & 1 \end{bmatrix}$$

The determinant of the matrix $\mathbf{A}$ can be found by multiplying the entries along diagonals, thus

$$\det \mathbf{A} = -2 - 4 + 4 + 6 = 4$$

Alternatively we could obtain it by expansion about the first row, recognising that the cofactors for the first row are the first column in the adjoint matrix. Thus

$$\det \mathbf{A} = 1 \times 4 + 2 \times 2 - 4 \times 1 = 4$$

The inverse matrix can now be obtained by the use of equation [9],

$$\mathbf{A}^{-1} = \frac{\text{adj } \mathbf{A}}{\det \mathbf{A}} = \frac{1}{4} \begin{bmatrix} 4 & 8 & 0 \\ -2 & -2 & 2 \\ 1 & -1 & -1 \end{bmatrix}$$

We can check this answer by recognising that we must have $\mathbf{AA}^{-1} = \mathbf{I}$, i.e.

$$\begin{bmatrix} 1 & 2 & 4 \\ 0 & -1 & -2 \\ 1 & 3 & 2 \end{bmatrix} \frac{1}{4} \begin{bmatrix} 4 & 8 & 0 \\ -2 & -2 & 2 \\ 1 & -1 & -1 \end{bmatrix} = \begin{bmatrix} 1 & 0 & 0 \\ 0 & 1 & 0 \\ 0 & 0 & 1 \end{bmatrix}$$

Review problems

2   Determine the inverse matrices for each of the matrices given in review problem 1.

5.2.2 Solving simultaneous equations

As indicated in the opening section (5.1) to this chapter, a set of simultaneous equations such as

$$a_{11}x_1 + a_{12}x_2 = c_1$$

$$a_{21}x_1 + a_{22}x_2 = c_2$$

can be written in the form (equation [3]),

$$\mathbf{x} = \mathbf{A}^{-1}\mathbf{c}$$

and hence solved.

Example

Solve, by the use of the inverse matrix, the following set of simultaneous equations:

$$2x + 3y + z = 11, \quad x - y + 2z = 5, \quad x + y - z = 0$$

The matrix for the coefficients is

$$\mathbf{A} = \begin{bmatrix} 2 & 3 & 1 \\ 1 & -1 & 2 \\ 1 & 1 & -1 \end{bmatrix}$$

The cofactors are thus

$$A_{11} = + \begin{vmatrix} -1 & 2 \\ 1 & -1 \end{vmatrix} = 1 - 2 = -1$$

$$A_{12} = - \begin{vmatrix} 1 & 2 \\ 1 & -1 \end{vmatrix} = 1 + 2 = 3$$

$$A_{13} = + \begin{vmatrix} 1 & -1 \\ 1 & 1 \end{vmatrix} = 1 + 1 = 2$$

$$A_{21} = - \begin{vmatrix} 3 & 1 \\ 1 & -1 \end{vmatrix} = 3 + 1 = 4$$

$$A_{22} = + \begin{vmatrix} 2 & 1 \\ 1 & -1 \end{vmatrix} = -2 - 1 = -3$$

$$A_{23} = - \begin{vmatrix} 2 & 3 \\ 1 & 1 \end{vmatrix} = -2 + 3 = 1$$

$$A_{31} = + \begin{vmatrix} 3 & 1 \\ -1 & 2 \end{vmatrix} = 6 + 1 = 7$$

$$A_{32} = - \begin{vmatrix} 2 & 1 \\ 1 & 2 \end{vmatrix} = -4 + 1 = -3$$

$$A_{33} = + \begin{vmatrix} 2 & 3 \\ 1 & -1 \end{vmatrix} = -2 - 3 = -5$$

Thus the cofactor matrix is

$$\begin{bmatrix} -1 & 3 & 2 \\ 4 & -3 & 1 \\ 7 & -3 & -5 \end{bmatrix}$$

The adjoint matrix is thus

$$\text{adj } \mathbf{A} = \begin{bmatrix} -1 & 4 & 7 \\ 3 & -3 & -3 \\ 2 & 1 & -5 \end{bmatrix}$$

The inverse of matrix $\mathbf{A}$ is given by equation [9] as

$$\mathbf{A}^{-1} = \frac{\text{adj } \mathbf{A}}{\det \mathbf{A}}$$

Using multiplication along the diagonals, the determinant of $\mathbf{A}$ is

$$\det \mathbf{A} = 2 + 6 + 1 + 1 - 4 + 3 = 9$$

Thus

$$A^{-1} = \frac{1}{9} \begin{bmatrix} -1 & 4 & 7 \\ 3 & -3 & -3 \\ 2 & 1 & -5 \end{bmatrix}$$

The simultaneous equations can be written as

$$\begin{bmatrix} x \\ y \\ z \end{bmatrix} = A^{-1} \begin{bmatrix} 11 \\ 5 \\ 0 \end{bmatrix}$$

Hence

$$\begin{bmatrix} x \\ y \\ z \end{bmatrix} = \frac{1}{9} \begin{bmatrix} -1 & 4 & 7 \\ 3 & -3 & -3 \\ 2 & 1 & -5 \end{bmatrix} \begin{bmatrix} 11 \\ 5 \\ 0 \end{bmatrix}$$

Thus

$$x = \frac{1}{9}(-11 + 20 + 0) = 1$$

$$y = \frac{1}{9}(33 - 15 + 0) = 2$$

$$z = \frac{1}{9}(22 + 5 + 0) = 3$$

Review problems

3 Solve, by determining the inverse matrices, the following sets of simultaneous equations:

(a) $2x + y + z = 1, \quad x + 3y - z = 0, \quad x - y + 2z = 4,$

(b) $2x + 3y + z = 0, \quad 3x - y - z = 8, \quad x + 2y + 3z = -3,$

(c) $2x + y + z = 7, \quad -x + 2y + 3z = -1, \quad 3x - y + z = 8,$

(d) $2x + y - z = 1, \quad x + 2y + z = 8, \quad x - y + 2z = 5,$

(e) $x + 2y + z = 5, \quad x - y - z = 0, \quad y + 2z = 3$

5.3 Cramer's rule

Consider a set of simultaneous equations

$$a_{11}x_1 + a_{12}x_2 + a_{13}x_3 = c_1$$

$$a_{21}x_1 + a_{22}x_2 + a_{23}x_3 = c_2$$

$$a_{31}x_1 + a_{32}x_2 + a_{33}x_3 = c_3$$

We can describe the relationship as (equation [3])

$$\mathbf{x} = \mathbf{A}^{-1}\mathbf{c}$$

Since the inverse matrix $\mathbf{A}^{-1}$ can be written as (equation [9])

$$\mathbf{A}^{-1} = \frac{\text{adj } \mathbf{A}}{\det \mathbf{A}}$$

then we have

$$\mathbf{x} = \frac{\text{adj } \mathbf{A}}{\det \mathbf{A}}\mathbf{c}$$

$$= \frac{1}{\det \mathbf{A}} \begin{bmatrix} A_{11} & A_{21} & A_{31} \\ A_{12} & A_{22} & A_{32} \\ A_{13} & A_{23} & A_{33} \end{bmatrix} \begin{bmatrix} c_1 \\ c_2 \\ c_3 \end{bmatrix}$$

Multiplying these two matrices gives

$$\mathbf{x} = \frac{1}{\det \mathbf{A}} \begin{bmatrix} c_1A_{11} + c_2A_{21} + c_3A_{31} \\ c_1A_{12} + c_2A_{22} + c_3A_{32} \\ c_1A_{13} + c_2A_{23} + c_3A_{33} \end{bmatrix}$$

Hence

$$x_1 = \frac{c_1A_{11} + c_2A_{21} + c_3A_{31}}{\det A} \qquad\qquad [10]$$

$$x_2 = \frac{c_1A_{12} + c_2A_{22} + c_3A_{32}}{\det A} \qquad\qquad [11]$$

$$x_3 = \frac{c_1A_{13} + c_2A_{23} + c_3A_{33}}{\det A} \qquad\qquad [12]$$

The numerators in equations [10], [11] and [12] represent the expansions of determinants by cofactors. Thus for the numerator in equation [10] we have

$$A_{11} = \begin{vmatrix} a_{22} & a_{23} \\ a_{32} & a_{33} \end{vmatrix}$$

$$A_{21} = -\begin{vmatrix} a_{12} & a_{13} \\ a_{32} & a_{33} \end{vmatrix}$$

$$A_{31} = \begin{vmatrix} a_{12} & a_{13} \\ a_{22} & a_{23} \end{vmatrix}$$

and so we can write

$$c_1 A_{11} + c_2 A_{21} + c_3 A_{31} = \begin{vmatrix} c_1 & a_{12} & a_{13} \\ c_2 & a_{22} & a_{23} \\ c_3 & a_{32} & a_{33} \end{vmatrix}$$

Hence

$$x_1 = \frac{\begin{vmatrix} c_1 & a_{12} & a_{13} \\ c_2 & a_{22} & a_{23} \\ c_3 & a_{32} & a_{33} \end{vmatrix}}{\det \mathbf{A}} \qquad [13]$$

In a similar way we can write

$$x_2 = \frac{\begin{vmatrix} a_{11} & c_1 & a_{13} \\ a_{21} & c_2 & a_{23} \\ a_{31} & c_3 & a_{33} \end{vmatrix}}{\det \mathbf{A}} \qquad [14]$$

$$x_3 = \frac{\begin{vmatrix} a_{11} & a_{12} & c_1 \\ a_{21} & a_{22} & c_2 \\ a_{31} & a_{32} & c_3 \end{vmatrix}}{\det \mathbf{A}} \qquad [15]$$

In the numerator of each of the above equations appears the same determinant, that of $\mathbf{A}$, but modified by the replacement of the relevant column with the entries from the $\mathbf{c}$ matrix. The denominators of each equation are the same, namely $\mathbf{A}$. This result is known as *Cramer's rule*.

It is quite common to write equations [13], [14] and [15] in the form

$$x_1 = \frac{\det \mathbf{A}_1}{\det \mathbf{A}} \qquad [16]$$

$$x_2 = \frac{\det A_2}{\det A} \qquad [17]$$

$$x_3 = \frac{\det A_3}{\det A} \qquad [18]$$

where $\det A_1$, $\det A_2$ and $\det A_3$ represent the determinants in the numerators of equations [13], [14] and [15]. Hence

$$\frac{x_1}{\det A_1} = \frac{x_2}{\det A_2} = \frac{x_3}{\det A_3} = \frac{1}{\det A} \qquad [19]$$

Note that in some textbooks, the determinants of the numerators of equations [13], [14] and [15] are written in a different form, namely as

$$\begin{vmatrix} a_{11} & a_{12} & a_{13} & c_1 \\ a_{21} & a_{22} & a_{23} & c_2 \\ a_{31} & a_{32} & a_{33} & c_3 \end{vmatrix}$$

with the instruction that $\det A_1$ is to be obtained by covering up the 1 column, $\det A_2$ by covering up the 2 column, and $\det A_3$ by covering up the 3 column. With this definition we have

$$x_1 = \frac{\det A_1}{\det A} \qquad [20]$$

$$x_2 = -\frac{\det A_2}{\det A} \qquad [21]$$

$$x_3 = \frac{\det A_3}{\det A} \qquad [22]$$

$$\frac{x_1}{\det A_1} = -\frac{x_2}{\det A_2} = \frac{x_3}{\det A_3} = -\frac{1}{\det A} \qquad [23]$$

The minus signs appear in these equations because rows have been moved to different positions when compared with the previous definitions (see section 4.3.2).

Example

Solve, using Cramer's rule, the following simultaneous equations:

$$2x + 3y + z = 11, \quad x - y + 2z = 5, \quad x + y - z = 0$$

The determinant for the coefficients is

$$\det \mathbf{A} = \begin{vmatrix} 2 & 3 & 1 \\ 1 & -1 & 2 \\ 1 & 1 & -1 \end{vmatrix} = 2 + 6 + 1 + 1 - 4 + 3 = 9$$

Det $\mathbf{A}_1$ is obtained by replacing column 1 by the entries 11, 5 and 0. Hence

$$\det \mathbf{A}_1 = \begin{vmatrix} 11 & 3 & 1 \\ 5 & -1 & 2 \\ 0 & 1 & -1 \end{vmatrix} = 11 + 5 - 22 + 15 = 9$$

Hence, using equation [16],

$$x_1 = \frac{\det \mathbf{A}_1}{\det \mathbf{A}} = \frac{9}{9} = 1$$

Det $\mathbf{A}_2$ is obtained by replacing column 2 by the entries 11, 5 and 0. Hence

$$\det \mathbf{A}_2 = \begin{vmatrix} 2 & 11 & 1 \\ 1 & 5 & 2 \\ 1 & 0 & -1 \end{vmatrix} = -10 + 22 - 5 + 11 = 18$$

Hence, using equation [17],

$$x_2 = \frac{\det \mathbf{A}_2}{\det \mathbf{A}} = \frac{18}{9} = 2$$

Det $\mathbf{A}_3$ is obtained by replacing column 3 by the entries 11, 5 and 0. Hence

$$\det \mathbf{A}_3 = \begin{vmatrix} 2 & 3 & 11 \\ 1 & -1 & 5 \\ 1 & 1 & 0 \end{vmatrix} = 15 + 11 + 11 - 10 = 27$$

Hence, using equation [18],

$$x_3 = \frac{\det \mathbf{A}_3}{\det \mathbf{A}} = \frac{27}{9} = 3$$

This was the example considered in the previous section; you might like to compare the effort involved in obtaining the solutions in each case.

Review problems

4   Use Cramer's rule to solve the following sets of simultaneous equations:

(a) $x + 2y = 5$, $2x + y = 4$,

(b) $x - 2y = 6$, $2x + y = 2$,

(c) $3x + 4y = 5$, $x - 2y = 5$

5   Use Cramer's rule to solve the sets of three simultaneous equations given in review problem 3.

6   Use Cramer's rule to solve the following sets of four simultaneous equations:

(a) $2x_1 + x_2 - x_3 + 2x_4 = 1$, $x_1 - 2x_2 + x_3 - x_4 = 4$,

$3x_2 + 2x_3 + x_4 = 2$, $x_1 + x_2 + 2x_4 = 2$

(b) $x_1 + x_2 + x_3 + x_4 = 2$, $x_1 - x_2 + 2x_3 + x_4 = -1$,

$2x_2 + x_3 + 2x_4 = 1$, $x_1 + 3x_2 + x_4 = 5$

7   Determine the cofactor matrix and the adjoint matrix for the following matrices:

(a) $\begin{bmatrix} 1 & 0 & 0 \\ 0 & 2 & 0 \\ 0 & 0 & 3 \end{bmatrix}$, (b) $\begin{bmatrix} 1 & 0 & 0 \\ 2 & 1 & 0 \\ 3 & 0 & 1 \end{bmatrix}$, (c) $\begin{bmatrix} 3 & 2 & 1 \\ 3 & 2 & 2 \\ 1 & 3 & 1 \end{bmatrix}$

8   Determine the inverse matrices for each of the matrices in problem 7.

9   Determine the solutions of the following sets of simultaneous equations by (a) determining the inverse matrices, (b) using Cramer's rule:

(a) $x - y - z = 5$, $2x + y + 2z = 5$, $x + 2y + 3z = 0$,

(b) $2x + y - z = -2$, $x - y + 2z = 1$, $3x + 2y + z = 3$,

(c) $x + 2y + z = 6$, $2x - y - z = 1$, $-x + 3y + 2z = 5$,

(d) $2x + y - z = 3$, $x - y + 2z = 3$, $3x + 2y + z = 14$

Further problems

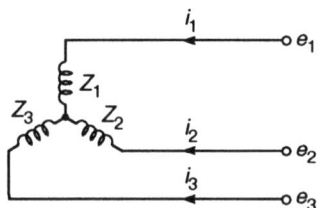

Fig. 5.1 Problem 10

10 The currents i_1, i_2 and i_3 in a star-connection (figure 5.1) of a three-phase supply are given by the equations

$$Z_1 i_1 - Z_2 i_2 = e_1 - e_2$$

$$Z_2 i_2 - Z_3 i_3 = e_2 - e_3$$

$$i_1 + i_2 + i_3 = 0$$

Z_1, Z_2 and Z_3 are impedances and e_1, e_2 and e_3 the phase e.m.f.s. Determine the three currents.

11 The velocity v of a car after a time t when accelerating from an initial velocity u at $t = 0$ with a constant acceleration a is given by $v = u + at$. If $v = 20$ m/s when $t = 5$ s and $v = 26$ m/s when $t = 7$ s, determine, using Cramer's rule, the values of u and a.

12 The equation of a straight line is $ax + by = c$. Determine the values of a, b and c if the line passes through the points (1, 2) and (4, −3).
Hint: solve the equation for the two ratios a/c and b/c.

6 Circuit analysis

6.1 Circuit analysis

Consider a simple circuit involving a d.c. voltage V applied across a circuit element of resistance R. If the resistance obeys Ohm's law then the steady-state current I is given by

$$V = RI$$

We thus have a linear relationship (see section 1.1) between the voltage and current variables. A circuit element that has a linear voltage–current relationship is said to be a *linear element*. When Kirchhoff's laws are applied to circuits with linear elements, linear equations relating the variables are produced. If such circuits have branches then we obtain sets of simultaneous linear equations.

This chapter is intended to illustrate how linear equations can be derived for such circuits and how matrices and determinants can be used in the solution of such equations.

6.1.1 Kirchhoff's laws

Kirchhoff's current law can be stated as: at any junction in an electrical circuit the current entering the junction equals the current leaving it. This is often stated in the form: the algebraic sums of the currents at each node in a circuit must equal zero. The term *node* is used for the point at which two or more conductors are joined to form a junction and the algebraic sum means that the direction of the current at a junction must be taken into account. We might take the current entering the junction to be positive and that leaving it to be negative. Thus, for the circuit shown in figure 6.1, there is a node at point 1 with the current I_A being positive and I_B and I_C negative. Hence Kirchhoff's current law gives for node *1*

$$I_A - I_B - I_C = 0$$

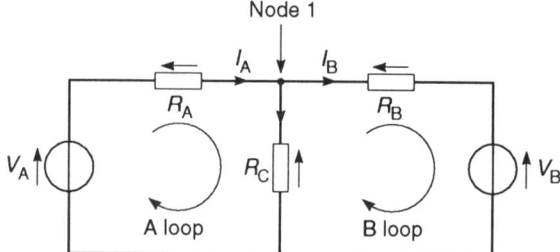

Fig. 6.1 Application of Kirchhoff's laws

Kirchhoff's voltage law can be stated as: around any closed path in a circuit the algebraic sum of the voltages across the components and the algebraic sum of the applied voltages must be zero. A closed path, often called a loop, is a path formed by selecting a starting node, proceeding through the circuit elements in that path, and returning to the starting node without going through a circuit element more than once. Thus for the circuit shown in figure 6.1, if we start at node 1 then we can have a closed path by proceeding through R_C, V_A and R_A back to node 1, or through R_B, V_B and R_C back to node 1, or through R_B, V_B, V_A and R_A back to node 1. The algebraic sum means that if we proceed through a circuit element in the direction of a voltage rise then the voltage is taken to be positive, if we proceed through the element in the direction of a voltage drop it is negative. The positive directions of the voltages are often indicated by arrows. The voltage rise across a circuit element is in the opposite direction to that of the current through it (think of the current flowing downhill). For a voltage source, the positive direction of the voltage is taken as being from the negative pole to the positive pole of the source and is generally indicated in a circuit diagram by an arrow in that direction. Thus for the circuit in figure 6.1 we can write, for the three circuit loops when we start from node 1 and move clockwise round the loops,

$$ -R_A I_A - R_C I_C + V_A = 0 $$

$$ -R_B I_B - V_B + R_C I_C = 0 $$

$$ -R_B I_B - V_B + V_A - R_A I_A = 0 $$

Kirchhoff's laws are really just statements of the principles of conservation of charge and conservation of energy. Thus the current law is just stating that the rate at which charge reaches a junction in a circuit is equal to the rate at which it leaves. The voltage law is just stating that the sum of the energy dissipated in the elements in a circuit equals the energy supplied. These are linear relationships and so Kirchhoff's laws are linear.

Two circuit analysis techniques which involve the use of Kirchhoff's laws are node voltage analysis and mesh analysis.

Node voltage analysis (see section 6.2) uses Kirchhoff's current law, mesh analysis (see section 6.3) involves Kirchhoff's voltage law.

6.2 Node voltage analysis

Node voltage analysis is the term used to describe a technique of systematically applying Kirchhoff's current law to a circuit. What we are concerned with are the relationships between the currents at each principal node in the circuit. A principal node is one at which three or more circuit elements join. One of the principal nodes is taken as a reference node, all the voltages at the other principal nodes being considered with reference to this node.

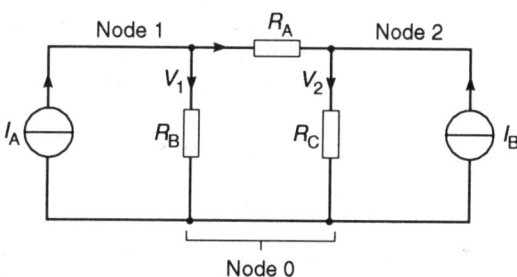

Fig. 6.2 Node voltage analysis

Consider the circuit shown in figure 6.2. This contains independent current sources, the term independent being used because the current does not depend on any circuit conditions. There are three principal nodes, 0, 1 and 2. The reference node is taken as node 0. Then the voltages at non-reference nodes are all taken relative to this reference node. These are taken to be V_1 at node 1 and V_2 at node 2. Thus the voltage across resistance R_A is thus $V_1 - V_2$, that across R_B is V_1, and across R_C is V_2. The current through R_A away from node 1 is $(V_1 - V_2)/R_A$, through R_B away from node 1 is V_1/R_B and through R_C away from node 2 is V_2/R_C.

Applying Kirchhoff's current law to node 1 gives

$$I_A = \frac{V_1 - V_2}{R_A} + \frac{V_1}{R_B}$$

Likewise, for node 2 we have

$$I_B = \frac{V_2 - V_1}{R_A} + \frac{V_2}{R_C}$$

If the circuit had a voltage source, rather than a current source, connected between two nodes, then we have to deal with the situation in a different way. If the voltage source is ideal, i.e. has zero internal resistance, then we can regard it as a short circuit with the voltage difference between the nodes being given by the value of the voltage. This is illustrated in the following examples. If the voltage source is not ideal we can replace it by its current equivalent.

By this means equations can be generated for each non-reference principal node. The number of simultaneous equations will be $(n - 1)$, where n is the number of principal nodes. The 1 is subtracted because one of the nodes is the reference node. With just two equations, straightforward elimination techniques can be used to obtain the variables (see chapter 1). When the analysis of a circuit yields more than two simultaneous equations, Gaussian elimination (see chapter 2), carrying out matrix operations (see chapters 3 and 5) and Cramer's rule (see chapter 5) are useful techniques that can be used to solve the equations. The following examples illustrate node voltage analysis of circuits using a range of methods for the solution of the equations. The first example involves only one equation, the second example two equations and the third example three equations.

The procedure for carrying out node analysis of a circuit can be summarised as:

1 Label the principal nodes on a circuit diagram, selecting one as the reference node.
2 If the circuit contains only current sources, apply Kirchhoff's current law to each non-reference principal node.
3 If the circuit contains voltage sources, consider each such source to be effectively a short circuit with the voltage value assigned to the appropriate node. If a voltage source is not ideal, convert it to an equivalent current source. Then apply Kirchhoff's current law to each non-reference principal node.
4 Solve the resulting simultaneous equations.

Example

For the circuit shown in figure 6.3, use node voltage analysis to obtain an equation relating the node voltages and hence obtain the current in each of the circuit branches.

Node 0 is selected as the reference node. The other principal node is node 1 and thus we are going to apply Kirchhoff's current law to that node. The circuit contains voltage sources. These are to be

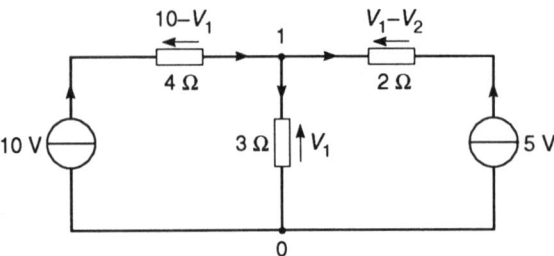

Fig. 6.3 Example

regarded as short circuits, but taking account of the voltages with regard to those at nodes. The voltage across the 4 Ω resistance is thus $(10 - V_1)$ and so the current through it, towards node 1, is given by $(10 - V_1)/4$. The voltage across the 3 Ω resistance is V_1 and so the current through it, away from node 1, is $V_1/3$. The voltage across the 2 Ω resistance is $(V_2 - 5)$ and so the current through it, away from node 1, is $(V_2 - 5)/2$. Hence Kirchhoff's current law gives

$$\frac{10 - V_1}{4} = \frac{V_1}{3} + \frac{V_1 - 5}{2}$$

Thus

$$30 - 3V_1 - 4V_1 - 6V_1 + 30 = 0$$

and $V_1 = 4.62$ V.

The current through the 4 Ω branch is thus given by $(10 - 4.62)/4 = 1.35$ A, through the 3 Ω branch $4.62/3 = 1.54$ A, and through the 2 Ω branch $(4.62 - 5)/2 = -0.19$ A. The minus sign is because the direction is opposite to that assumed in the figure.

Example

Determine the branch currents for the circuit shown in figure 6.4.

The circuit shows two constant current sources. The nodes have been marked 0, 1 and 2. We will take node 0 as the reference node. Arrows have been put alongside the resistance elements in an arbitrary way to indicate the directions being assumed for the voltages across them. Applying Kirchhoff's current law to node 1 we have

$$5 - \frac{V_1}{1} - \frac{V_1 - V_2}{2} = 0$$

$$10 - 2V_1 - V_1 + V_2 = 0$$

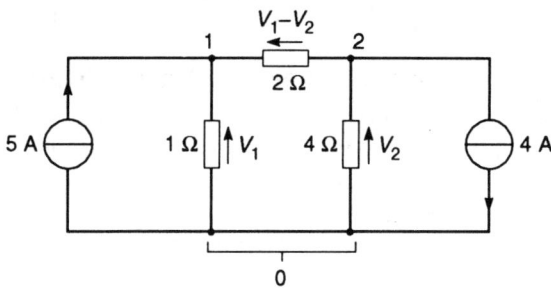

Fig. 6.4 Example

Thus for node 1

$$3V_1 - V_2 = 10 \qquad\qquad [1]$$

Applying Kirchhoff's current law to node 2 we have

$$\frac{V_1 - V_2}{2} - \frac{V_2}{4} - 4 = 0$$

$$2V_1 - 2V_2 - V_2 - 16 = 0$$

Thus for node 2

$$2V_1 - 3V_2 = 16 \qquad\qquad [2]$$

Equations [1] and [2] can be solved by using elimination (see chapter 1). Thus we can eliminate V_2 by subtracting equation [2] from three times equation [1]. Hence

$$7V_1 = 14$$

and so $V_1 = 2$ V. Back substitution of this value into equation [1] gives $V_2 = -4$ V.

As an alternative, we could have solved equations [1] and [2] by the use of Gaussian elimination (see chapter 2). Thus the augmented matrix for equations [1] and [2] is

$$\begin{bmatrix} 3 & -1 & 10 \\ 2 & -3 & 16 \end{bmatrix}$$

Dividing the first row by 3 gives

$$\begin{bmatrix} 1 & -\frac{1}{3} & \frac{10}{3} \\ 2 & -3 & 16 \end{bmatrix}$$

Subtracting twice the first row from the second row gives

$$\begin{bmatrix} 1 & -\frac{1}{3} & \frac{10}{3} \\ 0 & -\frac{7}{3} & \frac{28}{3} \end{bmatrix}$$

Dividing the second row by $-7/3$ gives

$$\begin{bmatrix} 1 & -\frac{1}{3} & \frac{10}{3} \\ 0 & 1 & -4 \end{bmatrix}$$

Thus $V_2 = -4$ V. We can obtain V_1 by back substitution into equation [1] or [2] or by adding a third of row 2 to row 1 to give

$$\begin{bmatrix} 1 & 0 & 2 \\ 0 & 1 & -4 \end{bmatrix}$$

Thus $V_1 = 2$ V.

As an alternative, we could have solved equations [1] and [2] by the use of matrix operations (see chapter 3). Thus the matrices for these equations are

$$\begin{bmatrix} 3 & -1 \\ 2 & -3 \end{bmatrix}\begin{bmatrix} V_1 \\ V_2 \end{bmatrix} = \begin{bmatrix} 10 \\ 16 \end{bmatrix}$$

i.e. $\mathbf{Av} = \mathbf{c}$. The coefficient matrix is thus

$$\mathbf{A} = \begin{bmatrix} 3 & -1 \\ 2 & -3 \end{bmatrix}$$

The inverse of the coefficient matrix, i.e. $\mathbf{A}^{-1}$, is given by equation [15] in chapter 2 for a matrix of the form

$$\begin{bmatrix} a & b \\ c & d \end{bmatrix}$$

as

$$\mathbf{A}^{-1} = \frac{1}{ad - bc}\begin{bmatrix} d & -b \\ -c & a \end{bmatrix}$$

Thus

$$\mathbf{A}^{-1} = \frac{1}{-9+2}\begin{bmatrix} -3 & 1 \\ -2 & 3 \end{bmatrix}$$

and so, since we can write $\mathbf{v} = \mathbf{A}^{-1}\mathbf{c}$, then

$$\begin{bmatrix} V_1 \\ V_2 \end{bmatrix} = \begin{bmatrix} \frac{3}{7} & -\frac{1}{7} \\ \frac{2}{7} & -\frac{3}{7} \end{bmatrix}\begin{bmatrix} 10 \\ 16 \end{bmatrix}$$

$$= \begin{bmatrix} \frac{30}{7} - \frac{16}{7} \\ \frac{20}{7} - \frac{48}{7} \end{bmatrix} = \begin{bmatrix} 2 \\ -4 \end{bmatrix}$$

Thus $V_1 = 2$ V and $V_2 = -4$ V.

An alternative way of solving the two equations is by the use of Cramer's rule (see section 5.3). Thus, equation [19] in chapter 5 gives

$$\frac{x_1}{\det \mathbf{A}_1} = \frac{x_2}{\det \mathbf{A}_2} = \frac{1}{\det \mathbf{A}}$$

where $\det \mathbf{A}_1$ is $\det \mathbf{A}$ with the first column replaced by the entries from the **c** matrix and $\det \mathbf{A}_2$ is with the second column replaced by the entries from the **c** matrix. Thus, using this rule

$$\frac{V_1}{\begin{vmatrix} 10 & -1 \\ 16 & -3 \end{vmatrix}} = \frac{V_2}{\begin{vmatrix} 3 & 10 \\ 2 & 16 \end{vmatrix}} = \frac{1}{\begin{vmatrix} 3 & -1 \\ 2 & -3 \end{vmatrix}}$$

Hence

$$\frac{V_1}{-30+16} = \frac{V_2}{48-20} = \frac{1}{-9+2}$$

Hence $V_1 = 2$ V and $V_2 = -4$ V.

There is another way of writing Cramer's rule (chapter 5, equation [23])

$$\frac{x_1}{\det \mathbf{A}_1} = -\frac{x_2}{\det \mathbf{A}_2} = -\frac{1}{\det \mathbf{A}}$$

in which the determinants of $\mathbf{A}_1$ and $\mathbf{A}_2$ are given by writing determinant $\mathbf{A}$ with the **c** matrix entries added as a final extra column, then covering up the first column to give the determinant of $\mathbf{A}_1$ and covering up the second column to give the determinant of $\mathbf{A}_2$. Thus, using this rule with equations [1] and [2], we have

$$\frac{V_1}{\begin{vmatrix} -1 & -10 \\ -3 & -16 \end{vmatrix}} = \frac{-V_2}{\begin{vmatrix} 3 & -10 \\ 2 & -16 \end{vmatrix}} = \frac{1}{\begin{vmatrix} 3 & -1 \\ 2 & -3 \end{vmatrix}}$$

$$\frac{V_1}{16-30} = \frac{-V_2}{-48+20} = \frac{1}{-9+2}$$

Hence $V_1 = 2$ V and $V_2 = -4$ V.

The current through the 1 Ω branch is thus $V_1/1 = 2/1 = 2$ A. The current through the 4 Ω branch is $V_2/4 = -4/4 = -1$ A. The current through the 2 Ω branch is $(V_1 - V_2)/2 = (2 + 4)/2 = 3$ A.

Example

Determine the currents through the branches of the circuit shown in figure 6.5.

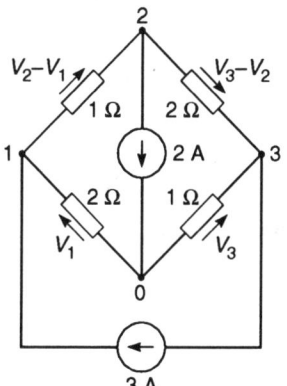

$V_2 - V_1$

$V_3 - V_2$

$1\,\Omega$ $2\,\Omega$

2 A

$2\,\Omega$ $1\,\Omega$

V_1 V_3

3 A

Fig. 6.5 Example

Taking node 0 as the reference node and assuming arbitrary directions for the voltages across the various circuit elements, the following equations are then obtained as a result of applying node voltage analysis to nodes 1, 2 and 3. For node 1 we have

$$3 - \frac{V_1}{2} + \frac{V_2 - V_1}{1} = 0$$

$$3V_1 - 2V_2 = 6 \qquad [3]$$

For node 2 we have

$$-\frac{V_2 - V_1}{1} + \frac{V_3 - V_2}{2} - 2 = 0$$

$$2V_1 - 3V_2 + V_3 = 4 \qquad [4]$$

For node 3 we have

$$-\frac{V_3 - V_2}{2} - \frac{V_3}{1} - 3 = 0$$

$$V_2 - 3V_3 = 6 \qquad [5]$$

The three variables can be found by means of Gaussian elimination (see chapter 1). The augmented matrix for equations [3], [4] and [5] is

$$\begin{bmatrix} 3 & -2 & 0 & 6 \\ 2 & -3 & 1 & 4 \\ 0 & 1 & -3 & 6 \end{bmatrix}$$

Dividing the first row by 3 gives

$$\begin{bmatrix} 1 & -\frac{2}{3} & 0 & 2 \\ 2 & -3 & 1 & 4 \\ 0 & 1 & -3 & 6 \end{bmatrix}$$

Subtracting twice the first row from the second row gives

$$\begin{bmatrix} 1 & -\frac{2}{3} & 0 & 2 \\ 0 & -\frac{5}{3} & 1 & 0 \\ 0 & 1 & -3 & 6 \end{bmatrix}$$

Dividing the second row by $-5/3$ gives

$$\begin{bmatrix} 1 & -\frac{2}{3} & 0 & 2 \\ 0 & 1 & -\frac{3}{5} & 0 \\ 0 & 1 & -3 & 6 \end{bmatrix}$$

Subtracting the second row from the third row gives

$$\begin{bmatrix} 1 & -\frac{2}{3} & 0 & 2 \\ 0 & 1 & -\frac{3}{5} & 0 \\ 0 & 0 & -\frac{12}{5} & 6 \end{bmatrix}$$

Dividing the third row by $-12/5$ gives

$$\begin{bmatrix} 1 & -\frac{2}{3} & 0 & 2 \\ 0 & 1 & -\frac{3}{5} & 0 \\ 0 & 0 & 1 & -\frac{5}{2} \end{bmatrix}$$

Thus $V_3 = -2.5$ V. We can obtain the values of the other variables by back substitution or by operations on the rows in the matrix. Adding 3/5 of the third row to the second row gives

$$\begin{bmatrix} 1 & -\frac{2}{3} & 0 & 2 \\ 0 & 1 & 0 & -\frac{3}{2} \\ 0 & 0 & 1 & -\frac{5}{2} \end{bmatrix}$$

Thus $V_2 = -1.5$ V. Adding 2/3 of the second row to the first row gives

$$\begin{bmatrix} 1 & 0 & 0 & 1 \\ 0 & 1 & 0 & -\frac{3}{2} \\ 0 & 0 & 1 & -\frac{5}{2} \end{bmatrix}$$

Thus $V_1 = 1$ V.

Alternatively we could have solved equations [3], [4] and [5] by writing matrices in the following form and then obtaining the inverse matrix.

$$\begin{bmatrix} 3 & -2 & 0 \\ 2 & -3 & 1 \\ 0 & 1 & -3 \end{bmatrix} \begin{bmatrix} V_1 \\ V_2 \\ V_3 \end{bmatrix} = \begin{bmatrix} 6 \\ 4 \\ 6 \end{bmatrix}$$

i.e. $\mathbf{Av} = \mathbf{c}$ and so $\mathbf{v} = \mathbf{A}^{-1}\mathbf{c}$. The inverse matrix $\mathbf{A}^{-1}$ could be found by row operations (see section 3.6.2) or by determinants (see section 5.2.1). Here we will illustrate its determination by the use of determinants. Equation [9], chapter 5, gives

$$\mathbf{A}^{-1} = \frac{\text{adj } \mathbf{A}}{\det \mathbf{A}}$$

where adj $\mathbf{A}$ is the transpose of the matrix of the cofactors of matrix $\mathbf{A}$. The cofactors are

$$A_{11} = + \begin{vmatrix} -3 & 1 \\ 1 & -3 \end{vmatrix} = 9 - 1 = 8$$

$$A_{12} = - \begin{vmatrix} 2 & 1 \\ 0 & -3 \end{vmatrix} = 6$$

$$A_{13} = + \begin{vmatrix} 2 & -3 \\ 0 & 1 \end{vmatrix} = 2$$

$$A_{21} = - \begin{vmatrix} -2 & 0 \\ 1 & -3 \end{vmatrix} = -6$$

$$A_{22} = + \begin{vmatrix} 3 & 0 \\ 0 & -3 \end{vmatrix} = -9$$

$$A_{23} = - \begin{vmatrix} 3 & -2 \\ 0 & 1 \end{vmatrix} = -3$$

$$A_{31} = + \begin{vmatrix} -2 & 0 \\ -3 & 1 \end{vmatrix} = -2$$

$$A_{32} = - \begin{vmatrix} 3 & 0 \\ 2 & 1 \end{vmatrix} = -3$$

$$A_{33} = + \begin{vmatrix} 3 & -2 \\ 2 & -3 \end{vmatrix} = -9 + 4 = -5$$

Thus the matrix of the cofactors is

$$\begin{bmatrix} 8 & 6 & 2 \\ -6 & -9 & -3 \\ -2 & -3 & -5 \end{bmatrix}$$

and hence

$$\text{adj } \mathbf{A} = \begin{bmatrix} 8 & -6 & -2 \\ 6 & -9 & -3 \\ 2 & -3 & -5 \end{bmatrix}$$

Thus

$$\mathbf{A}^{-1} = \frac{\begin{bmatrix} 8 & -6 & -2 \\ 6 & -9 & -3 \\ 2 & -3 & -5 \end{bmatrix}}{\begin{vmatrix} 3 & -2 & 0 \\ 2 & -3 & 1 \\ 0 & 1 & -3 \end{vmatrix}} = \frac{1}{12} \begin{bmatrix} 8 & -6 & -2 \\ 6 & -9 & -3 \\ 2 & -3 & -5 \end{bmatrix}$$

Hence

$$\begin{bmatrix} V_1 \\ V_2 \\ V_3 \end{bmatrix} = \frac{1}{12} \begin{bmatrix} 8 & -6 & -2 \\ 6 & -9 & -3 \\ 2 & -3 & -5 \end{bmatrix} \begin{bmatrix} 6 \\ 4 \\ 6 \end{bmatrix}$$

$$V_1 = \frac{1}{12}(48 - 24 - 12) = 1 \text{ V}$$

$$V_2 = \frac{1}{12}(36 - 36 - 18) = -1.5 \text{ V}$$

$$V_3 = \frac{1}{12}(12 - 12 - 30) = -2.5 \text{ V}$$

Alternatively we could have solved the three equations by the use of Cramer's rule. Thus, using the form given by equation [19] in chapter 5,

$$\frac{x_1}{\det \mathbf{A}_1} = \frac{x_2}{\det \mathbf{A}_2} = \frac{x_3}{\det \mathbf{A}_3} = \frac{1}{\det \mathbf{A}}$$

we have

$$V_1 = \frac{\det A_1}{\det A} = \frac{\begin{vmatrix} 6 & -2 & 0 \\ 4 & -3 & 1 \\ 6 & 1 & -3 \end{vmatrix}}{\begin{vmatrix} 3 & -2 & 0 \\ 2 & -3 & 1 \\ 0 & 1 & -3 \end{vmatrix}} = \frac{54 - 12 - 6 - 24}{27 - 3 - 12} = 1 \text{ V}$$

$$V_2 = \frac{\det A_2}{\det A} = \frac{\begin{vmatrix} 3 & 6 & 0 \\ 2 & 4 & 1 \\ 0 & 6 & -3 \end{vmatrix}}{\begin{vmatrix} 3 & -2 & 0 \\ 2 & -3 & 1 \\ 0 & 1 & -3 \end{vmatrix}} = \frac{-36 - 18 + 36}{27 - 3 - 12} = -1.5 \text{ V}$$

$$V_3 = \frac{\det A_3}{\det A} = \frac{\begin{vmatrix} 3 & -2 & 6 \\ 2 & -3 & 4 \\ 0 & 1 & 6 \end{vmatrix}}{\begin{vmatrix} 3 & -2 & 0 \\ 2 & -3 & 1 \\ 0 & 1 & -3 \end{vmatrix}} = \frac{-54 + 12 - 12 + 24}{27 - 3 - 12}$$

$$= -2.5 \text{ V}$$

The current through the branch between nodes 1 and 2 is thus $(V_2 - V_1)/1 = -2.5$ A. The current through the branch between nodes 2 and 3 is $(V_3 - V_2)/2 = 0.5$ A. The current through the branch between nodes 1 and 0 is $V_1/2 = 0.5$ A. The current through the branch between nodes 3 and 0 is $V_3/1 = -2.5$ A.

Review problems

1   Determine the node voltages for the following sets of equations:

(a) $2V_1 - V_2 = 7$, $V_1 - 3V_2 = 1$,

(b) $2V_1 - V_2 = 5$, $V_1 - 2V_2 + V_3 = 5$, $V_2 - 2V_3 = -3$,

(c) $V_1 - V_2 = 3$, $V_1 + 2V_2 - 3V_3 = 6$, $V_2 + 2V_3 = 4$,

(d) $V_1 - 2V_2 = 5$, $V_1 + 3V_2 - V_3 = -7$, $V_2 + 4V_3 - V_4 = 3$,

   $V_3 - V_4 = -1$

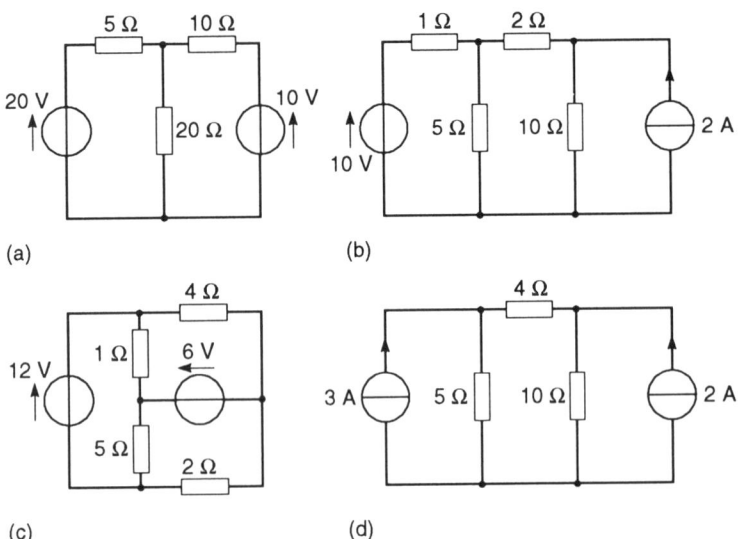

(a)

(b)

(c)

(d)

Fig. 6.6 Problem 2

2 Determine the currents in each branch of the circuits shown in figure 6.6.

6.2.1 General matrix method

The following is a more formal way of considering nodal analysis and is more suitable for computer implementation when circuits are complex.

Consider a circuit with n principal non-reference nodes and with only independent current sources. We will then consider just the kth node and assume that we can consider all the other nodes to be each connected to this node via some component (figure 6.7). Each component is represented by its conductance (note that conductance G is the reciprocal of resistance). If there is no such

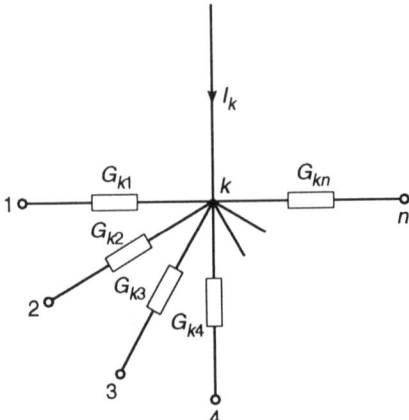

Fig. 6.7 Node k

component then all we have to do is consider its conductance to be zero. Applying Kirchhoff's current law to node k gives

$$(V_k - V_1)G_{k1} + (V_k - V_2)G_{k2} + (V_k - V_3)G_{k3} + ...$$
$$+ (V_k - V_n)G_{kn} = I_k$$

I_k is a current source to node k. If no such source exists then all we have to do is put the value equal to zero. All the voltage values are with reference to the reference node. We can rewrite this equation

$$V_1(- G_{k1}) + V_2(- G_{k2}) + V_3(- G_{k3}) + ... + V_k(G_{k1} + G_{k2}$$
$$+ G_{k3} + ... + G_{kn}) + ... + V_n(- G_{kn}) = I_k$$
[6]

Thus each node voltage, apart from V_k, is multiplied by minus the conductance between it and the kth node. V_k is multiplied by the sum of all the conductances entering that node.

We can write similar equations for all the other non-reference nodes. With n such nodes there will be n simultaneous equations. Thus for node 1, equation [6] becomes

$$V_1(G_{12} + G_{13} + G_{14} + ...) + V_2(- G_{12}) + V_3(- G_{13}) + ... = I_1$$

For node 2,

$$V_1(- G_{21}) + V_2(G_{21} + G_{23} + G_{24} + ...) + V_3(- G_{23}) + ... = I_2$$

If we write G_{11} for the sum of all the conductances terminating at node 1, G_{22} for the sum of all those terminating at node 2, etc. then we can represent the set of simultaneous equations, for all the non-reference nodes, as

$$\begin{bmatrix} G_{11} & -G_{12} & -G_{13} & ... \\ -G_{21} & G_{22} & -G_{23} & ... \\ -G_{31} & -G_{32} & G_{33} & ... \\ \vdots & \vdots & \vdots & \end{bmatrix} \begin{bmatrix} V_1 \\ V_2 \\ V_3 \\ \vdots \end{bmatrix} = \begin{bmatrix} I_1 \\ I_2 \\ I_3 \\ \vdots \end{bmatrix}$$
[7]

The square matrix is termed the *conductance matrix*. It is symmetric since $G_{jk} = G_{kj}$.

Example

Write the matrix equation for the node analysis of the circuit shown in figure 6.8.

The figure has the nodes numbered. For convenience the reference node has been labelled as node 0. This is because we do not write an equation for that node.

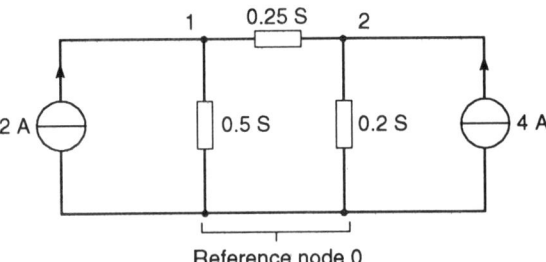

Fig. 6.8 Example

For node 1 we have an input current of 2 A. The sum of all conductances terminating at node 1 is $G_{11} = 0.5 + 0.25 = 0.75$ S and we have $G_{12} = 0.25$ S.

For node 2 we have an input current of 4 A. The sum of all conductances terminating at node 2 is $G_{22} = 0.25 + 0.2 = 0.45$ S and we have $G_{21} = 0.25$ S.

The matrix equation is thus

$$\begin{bmatrix} 0.75 & -0.25 \\ -0.25 & 0.45 \end{bmatrix} \begin{bmatrix} V_1 \\ V_2 \end{bmatrix} = \begin{bmatrix} 2 \\ 4 \end{bmatrix}$$

This matrix equation can be solved in one of the usual ways. This technique is particularly useful when the circuit is very complex and there are many nodes.

Review problems

3   Write the matrix equations for nodal analysis of the circuits shown in figure 6.9.

6.3 Mesh analysis

Mesh analysis is a systematic way of applying Kirchhoff's voltage law to circuits. It can only be applied to *planar circuits*, i.e. circuits which can be drawn on a plane surface. The term *closed path* or *loop* is used for a path that starts at a node and returns to the original node without passing through intermediate nodes more than once. A *mesh* is a special form of closed path, being one that does not contain any other closed paths within it. Thus for the circuit shown in figure 6.10, there are two meshes. For each mesh a mesh current is defined, such a current being considered to circulate round the mesh. The same direction of circulation must be chosen for all the mesh currents in a circuit. A usual convention is to make all the mesh currents circulate in a clockwise direction.

For the circuit shown in figure 6.10, the current through R_1 is the mesh current I_1. The current through R_2, which is common to the two meshes, is the algebraic sum of the two mesh currents, i.e. $I_1 - I_2$. The current through R_3 is the mesh current I_3. Thus if we apply Kirchhoff's voltage law to mesh 1 we have

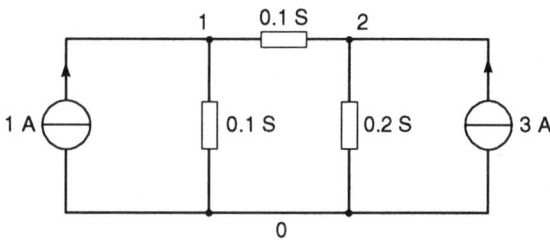

(a)

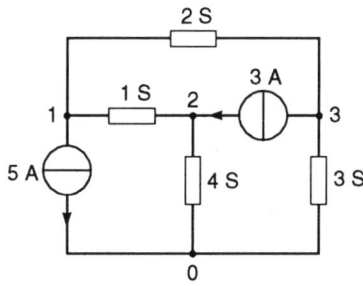

(b)

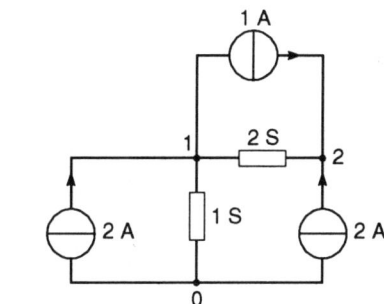

(c)

Fig. 6.9 Problem 3

Fig. 6.10 Mesh analysis

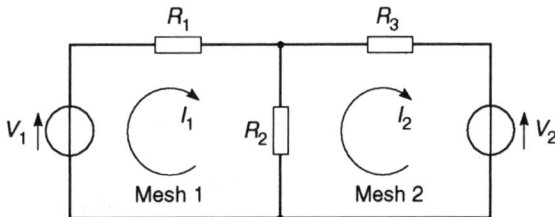

$$I_1 R_1 + (I_1 - I_2)R_2 = V_1$$

For mesh 2 we have

$$I_2 R_3 + (I_2 - I_1)R_2 = -V_2$$

The two simultaneous equations can then be solved to obtain the mesh currents and hence the currents through the individual components.

The general procedure for carrying out mesh analysis of a circuit can be summarised as:

1. Identify the meshes on the circuit diagram and assign a current to each, circulating in a clockwise direction.
2. If the circuit contains only voltage sources, apply Kirchhoff's voltage law to each mesh.
3. If the circuit contains current sources, replace each such source by an open circuit. The value of the current source can be directly related to the mesh currents to generate an equation. If the current source is not ideal, replace it by its equivalent voltage source.
4. Solve the resulting simultaneous equations.

Example

Determine, by mesh analysis, the currents in each of the branches of the circuit shown in figure 6.11.

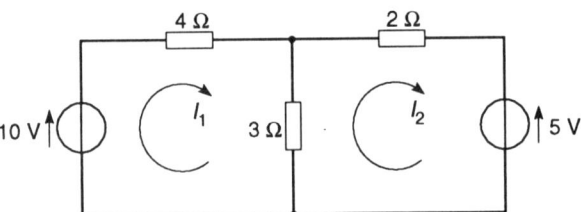

Fig. 6.11 Example

Applying Kirchhoff's voltage law to mesh 1 gives

$$4I_1 + 3(I_1 - I_2) = 10$$

and so

$$7I_1 - 3I_2 = 10 \qquad [8]$$

Applying Kirchhoff's voltage law to mesh 2 gives

$$2I_2 + 3(I_2 - I_1) = -5$$

and so

$$-3I_1 + 5I_2 = -5 \qquad [9]$$

The pair of simultaneous equations [8] and [9] can then be solved to give the mesh currents. As an illustration, Cramer's rule (see section 5.3) is used.

Thus, using equation [19] in chapter 5, we have

$$\frac{I_1}{\det \mathbf{A}_1} = \frac{I_2}{\det \mathbf{A}_2} = \frac{1}{\det \mathbf{A}}$$

$$\frac{I_1}{\begin{vmatrix} 10 & -3 \\ -5 & 5 \end{vmatrix}} = \frac{I_2}{\begin{vmatrix} 7 & 10 \\ -3 & -5 \end{vmatrix}} = \frac{1}{\begin{vmatrix} 7 & -3 \\ -3 & 5 \end{vmatrix}}$$

Thus

$$I_1 = \frac{50 - 15}{35 - 9} = 1.35 \text{ A}$$

$$I_2 = \frac{-35 + 30}{35 - 9} = -0.19 \text{ A}$$

The minus sign indicates that the current is in the opposite direction to the mesh current direction indicated in the figure. The current through the 4 Ω resistor is 1.35 A, through the 2 Ω resistor −0.19 A and through the 3 Ω resistor is the difference between the mesh currents, i.e. 1.35 − (−0.19) = 1.54 A.

Note that this example was also analysed, earlier in the chapter, by nodal analysis.

Example

Determine, by mesh analysis, the currents in the branches of the circuit shown in figure 6.12.

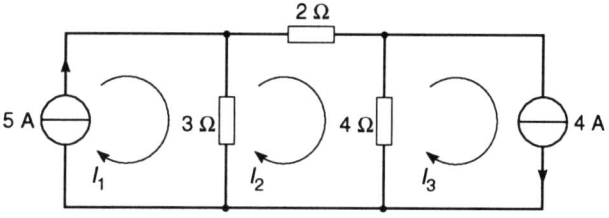

Fig. 6.12 Example

The three meshes and their currents are identified in the figure. It is readily apparent for mesh 1 that we must have $I_1 = 5$ A and for mesh 3 we have $I_3 = 4$ A. We cannot write any mesh equations for meshes 1 and 2. Each current source is regarded as ideal and as an open circuit. For mesh 2, we have

$$2I_2 + 4(I_2 - I_3) + 1(I_2 - I_1) = 0$$

which when reorganised gives

$$-1I_1 + 7I_2 - 4I_3 = 0$$

and hence $I_2 = \frac{1}{7}(5 + 4 \times 4) = 3$ A. The current through the 1 Ω resistor is thus $(I_1 - I_2) = 2$ A, through the 2 Ω resistor is $I_2 = 3$ A and through the 4 Ω resistor is $(I_3 - I_2) = 1$ A.

Note that this example was also analysed, earlier in the chapter, by nodal analysis.

Review problems

4 Determine, by mesh analysis, the currents in the circuits given in figure 6.6 for problem 2.

6.3.1 General matrix method

Consider the mesh equations in a circuit which contains only independent voltage sources and has n meshes. Figure 6.13 illustrates this, the resistors being labelled according to which two meshes they are common to, e.g. R_{12} is the resistor common to both mesh 1 and mesh 2.

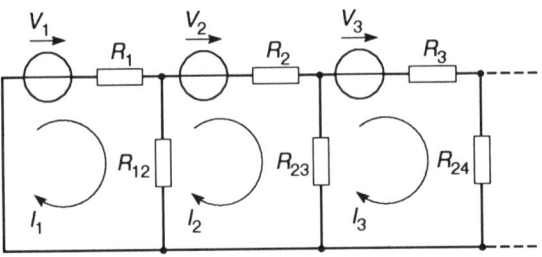

Fig. 6.13 Meshes

For mesh 1, we thus have

$$R_1 I_1 + R_{12}(I_1 - I_2) = V_1$$

$$(R_1 + R_{12})I_1 - R_{12}I_2 = V_1$$

The total resistance in mesh 1 is $R_1 + R_{12}$. If we designate this, so termed *self-resistance*, by R_{11} then the above equation becomes

$$R_{11}I_1 - R_{12}I_2 = V_1 \tag{10}$$

For mesh 2 we have

$$R_2 I_2 + R_{23}(I_2 - I_3) + R_{12}(I_2 - I_1) = V_2$$

$$-R_{12} I_1 + (R_2 + R_{12} + R_{23})I_2 - R_{23}I_3 = V_2$$

The total resistance in mesh 2 is $R_2 + R_{12} + R_{23}$. If we designate this self-resistance by R_{22} then the above equation becomes

$$-R_{12}I_1 + R_{22}I_2 - R_{23}I_3 = V_2 \qquad [11]$$

Equations can similarly be developed for all the meshes in the circuit. Equation [10] involved only an interaction between meshes 1 and 2, equation [11] an interaction between meshes 1, 2 and 3. The pattern indicated by equations [10] and [11] for a mesh in general can be summarised as: the product of the self-resistance of a mesh and its mesh current plus terms for each of the meshes with which the mesh concerned interacts. We can represent the set of equations given by this general equation, and equations [10] and [11], by

$$
\begin{bmatrix}
R_{11} & -R_{12} & -R_{13} & \cdots \\
-R_{21} & R_{22} & -R_{23} & \cdots \\
-R_{31} & -R_{32} & R_{33} & \cdots \\
\vdots & \vdots & \vdots &
\end{bmatrix}
\begin{bmatrix}
I_1 \\
I_2 \\
I_3 \\
\vdots
\end{bmatrix}
=
\begin{bmatrix}
V_1 \\
V_2 \\
V_3 \\
\vdots
\end{bmatrix}
\qquad [12]
$$

The square matrix is referred to as the *resistance matrix*.

Example

Write the matrix equation for mesh analysis for the circuit shown in figure 6.14.

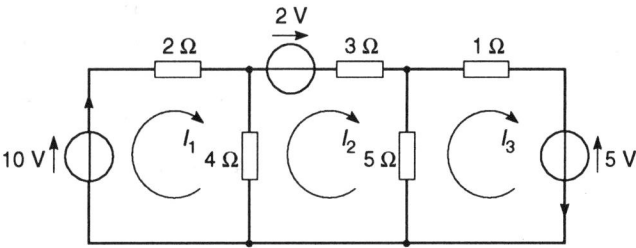

Fig. 6.14 Example

For mesh 1, the self-resistance is $R_{11} = 2 + 4 = 6 \ \Omega$. The only interaction between this mesh and other meshes is with mesh 2, for which $R_{12} = 4 \ \Omega$. Thus $R_{13} = 0$, since there is no such resistor common to both meshes. For mesh 1 we have $V_1 = 10$ V.

For mesh 2, the self-resistance is $R_{22} = 3 + 4 + 5 = 12 \ \Omega$. This mesh has resistors which are common with both meshes 1 and 3, giving $R_{21} = 4 \ \Omega$ and $R_{23} = 5 \ \Omega$. For mesh 2 we have $V_2 = 2$ V.

For mesh 3, the self-resistance is $R_{33} = 1 + 5 = 6 \ \Omega$. This mesh only has a resistor common with mesh 2, giving $R_{32} = 5 \ \Omega$. Thus $R_{31} = 0$, since there is no such resistor common to both meshes. For mesh 3 we have $V_3 = -5$ V.

The matrix equation is thus

$$
\begin{bmatrix} 6 & -4 & 0 \\ -4 & 12 & -5 \\ 0 & 5 & 6 \end{bmatrix} \begin{bmatrix} I_1 \\ I_2 \\ I_3 \end{bmatrix} = \begin{bmatrix} 10 \\ 2 \\ -5 \end{bmatrix}
$$

Review problems

5 Write the matrix equation for mesh analysis for the circuits shown in figure 6.15.

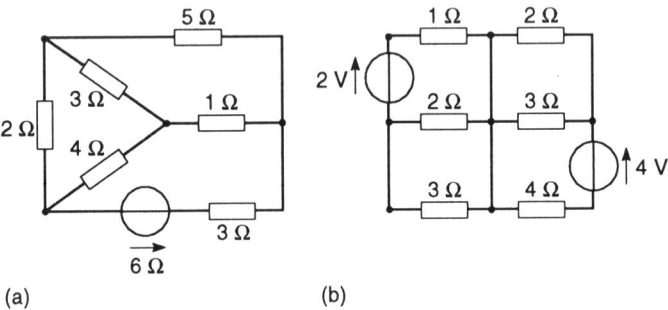

(a)   (b)

Fig. 6.15 Problem 5

Further problems

6 Determine, using (a) nodal analysis, (b) mesh analysis, (c) writing and then solving general matrix equations for nodal analysis, (d) writing and then solving general matrix equations for mesh analysis, the currents in each branch of the circuits shown in figure 6.16.
7 Write the matrix equation for nodal analysis of the circuits shown in figure 6.17.
8 Write the matrix equation for mesh analysis of the circuits shown in figure 6.18.

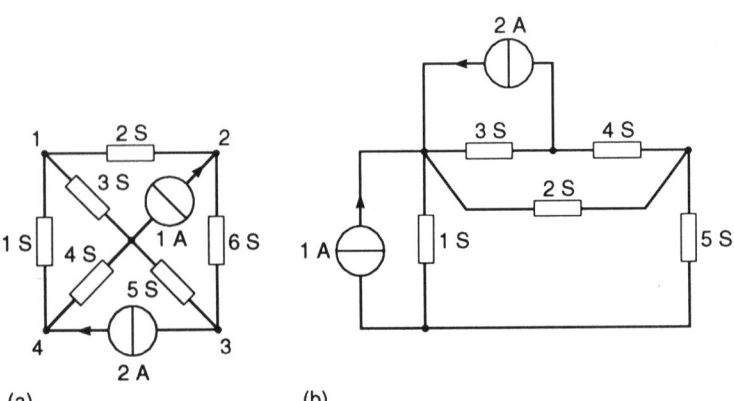

(a)   (b)

Fig. 6.17 Problem 7

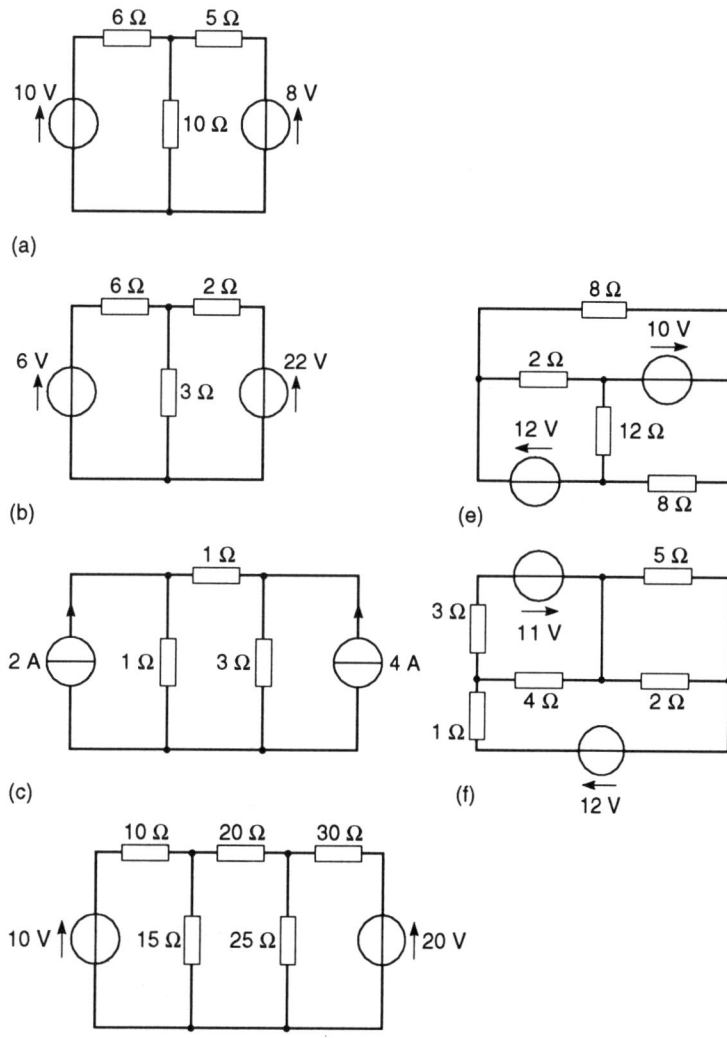

Fig. 6.16 Problem 6

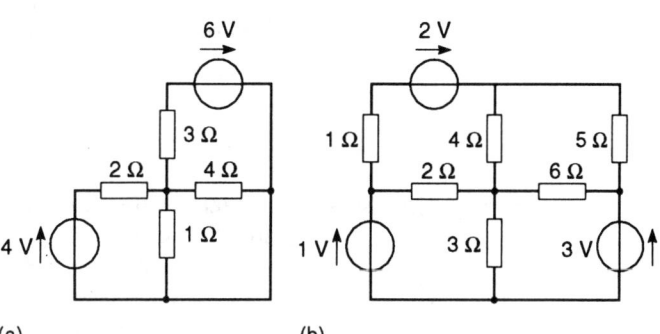

Fig. 6.18 Problem 8

7 Structural analysis

7.1 Matrix methods

Consider a simple spring with forces applied to its ends to stretch it. If the spring obeys Hooke's law then the change in length is proportional to the applied forces. The spring has a linear relationship between its extension x and the applied forces F. This can be represented by $F = kx$, where k is the *stiffness* of the spring. If we double the forces then we double the change in length. Also, if we apply a force F_1 and obtain an extension x_1, and a force F_2 gives an extension x_2, then a force of $F_1 + F_2$ will give an extension of $x_1 + x_2$. This is often termed the *principle of superposition*.

When forces are applied axially to the ends of a bar we can imagine the bar to be rather like a spring which is stretched or contracted, depending on the directions of the forces. Provided we do not exceed the limit of proportionality then we can consider the bar to be a linear element, like the spring described above.

This chapter is an illustration of the uses to which matrices can be put in structural analysis, being about the basic principles of the *matrix displacement method* for the analysis of structures. Such structures may be frameworks composed of many structural members. We can regard each member to be like a linear spring. The method can also be applied to two- or three-dimensional solids by imagining them to be structures made up of a large number of small linear elements like springs. This consideration of a solid in terms of a number of imaginary sub-elements is called the *finite element method*. The method is not restricted to just structural analysis but has much wider applications.

7.2 Analysis of spring elements

Consider a pin-jointed tie or strut for which we can assume that Hooke's law is obeyed. The points of attachment of the element to other parts of the structure are called nodes. We will take as a model for this member a spring. Figure 7.1 illustrates this with a spring of stiffness k between nodes 1 and 2.

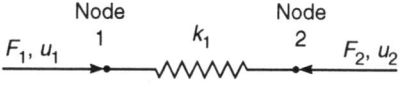

Fig. 7.1 Simple spring element

As a convention we will take the external force and displacement at a particular node to be denoted by F and u with a suffix indicating which node we are referring to. Forces and displacements are taken to be positive when in the positive x increasing direction, negative when in the opposite direction.

For the spring in figure 7.1, there is a displacement at one end of u_1 and a displacement, in the same direction, at the other end of u_2. The net extension of the spring is thus the difference between these two displacements. Since the stretched spring is in equilibrium, we must have no resultant force acting on the element, i.e.

$$F_1 + F_2 = 0$$

Thus $F_1 = -F_2$. Hence we have

$$F_1 = k(u_1 - u_2) = ku_1 - ku_2 \qquad [1]$$

$$F_2 = k(u_2 - u_1) = -ku_1 + ku_2 \qquad [2]$$

We can use matrices to describe the simultaneous equations [1] and [2]. Thus

$$\begin{bmatrix} k & -k \\ -k & k \end{bmatrix} \begin{bmatrix} u_1 \\ u_2 \end{bmatrix} = \begin{bmatrix} F_1 \\ F_2 \end{bmatrix} \qquad [3]$$

The square matrix is termed the *stiffness matrix* **K**. The column matrix of the displacements **u** is sometimes termed the *nodal displacement vector* and the column matrix of the forces **f** the *element force vector*.

$$\mathbf{Ku} = \mathbf{f} \qquad [4]$$

For a bar of material when subject to axial forces, within the Hooke's law region the stress, i.e. force F per unit cross-sectional area A, is proportional to the strain, i.e. extension x per unit length L. The constant of proportionality is the modulus of elasticity E. Thus, for a simple rod element (figure 7.2) we have

$$\frac{F}{A} = E\frac{x}{L}$$

$$F = \frac{EA}{L}x \qquad [5]$$

Thus the stiffness k is EA/L.

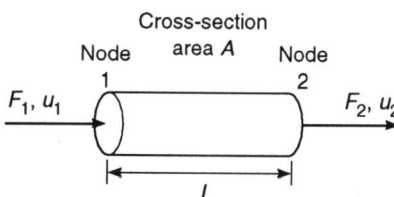

Fig. 7.2 Simple rod element

Example

What is the stiffness matrix for a rod of material of tensile modulus 200 GPa, length 1 m and cross-sectional area 0.01 m² when it is subject to axial forces?

The stiffness k is EA/L and thus

$$k = \frac{200 \times 10^9 \times 0.01}{1} = 2000 \text{ MN/m}$$

Hence the stiffness matrix is

$$\mathbf{K} = \begin{bmatrix} 2000 & -2000 \\ -2000 & 2000 \end{bmatrix} \text{MN/m}$$

Review problems

1 What is the stiffness matrix for a rod of material of tensile modulus 70 GPa, length 2 m and cross-sectional area 0.001 m² when subject to axial forces?

7.2.1 An assembly of in-line elements

Consider two in-line springs (figure 7.3). Here we have three nodes. We wish to consider each element in isolation, so we cut the structure at node 2 to obtain the free body equations for each element. For element 1 we then have

$$F_1 = k_1(u_1 - u_2) = k_1 u_1 - k_1 u_2 \qquad [6]$$

$$F_{12} = k_1(u_2 - u_1) = -k_1 u_1 + k_1 u_2 \qquad [7]$$

and can write

$$\begin{bmatrix} k_1 & -k_1 \\ -k_1 & k_1 \end{bmatrix} \begin{bmatrix} u_1 \\ u_2 \end{bmatrix} = \begin{bmatrix} F_1 \\ F_{12} \end{bmatrix} \qquad [8]$$

and for element 2

$$F_{22} = k_2(u_2 - u_3) = k_2 u_2 - k_2 u_3 \qquad [9]$$

$$F_3 = k_2(u_3 - u_2) = -k_2 u_2 + k_2 u_3 \qquad [10]$$

$$\begin{bmatrix} k_2 & -k_2 \\ -k_2 & k_2 \end{bmatrix} \begin{bmatrix} u_2 \\ u_3 \end{bmatrix} = \begin{bmatrix} F_{22} \\ F_3 \end{bmatrix} \qquad [11]$$

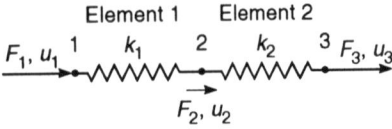

(a)

(b)

Fig. 7.3 Two in-line springs

As we have just three unknowns in terms of displacements u_1, u_2 and u_3 we could combine the equations to write three equations

$$F_1 = k_1 u_1 - k_1 u_2 \tag{12}$$

$$F_2 = F_{12} + F_{22} = -k_1 u_1 + (k_1 + k_2)u_2 - k_2 u_3 \tag{13}$$

$$F_3 = -k_2 u_2 + k_2 u_3 \tag{14}$$

These three equations can be written in matrix notation as

$$\begin{bmatrix} k_1 & -k_1 & 0 \\ -k_1 & k_1+k_2 & -k_2 \\ 0 & -k_2 & k_2 \end{bmatrix} \begin{bmatrix} u_1 \\ u_2 \\ u_3 \end{bmatrix} = \begin{bmatrix} F_1 \\ F_2 \\ F_3 \end{bmatrix} \tag{15}$$

Thus we have $\mathbf{Ku} = \mathbf{f}$, where $\mathbf{K}$ is the *stiffness matrix of the structure*, i.e. the assembly of the two in-line springs.

We could have derived equation [13] in another way. Equation [8] can be written as

$$\begin{bmatrix} k_1 & -k_1 & 0 \\ -k_1 & k_1 & 0 \\ 0 & 0 & 0 \end{bmatrix} \begin{bmatrix} u_1 \\ u_2 \\ u_3 \end{bmatrix} = \begin{bmatrix} F_1 \\ F_{12} \\ 0 \end{bmatrix}$$

and equation [11] as

$$\begin{bmatrix} 0 & 0 & 0 \\ 0 & k_2 & -k_2 \\ 0 & -k_2 & k_2 \end{bmatrix} \begin{bmatrix} u_1 \\ u_2 \\ u_3 \end{bmatrix} = \begin{bmatrix} 0 \\ F_{22} \\ F_3 \end{bmatrix}$$

If we add these two matrix equations we have

$$\begin{bmatrix} k_1 & -k_1 & 0 \\ -k_1 & k_1+k_2 & -k_2 \\ 0 & -k_2 & k_2 \end{bmatrix} \begin{bmatrix} u_1 \\ u_2 \\ u_3 \end{bmatrix} = \begin{bmatrix} F_1 \\ F_{12}+F_{22} \\ F_3 \end{bmatrix}$$

Hence we obtain the matrix equation given in equation [15].

Suppose we have more than two springs in-line. We can analyse the arrangement in the same way as with two springs. The basic rules that emerge for the writing of the stiffness matrix are:

1 The term in the stiffness matrix at location *jj* consists of the sum of the stiffnesses of the individual elements meeting at node *j*.
2 The term in the matrix at location *ij* consists of minus the stiffness of the element that is joining node *i* to *j*, or minus the

sum of the stiffnesses of all the elements that are joining node i to j.

All that the above rules are summarising are the results of adding the matrices for the individual elements.

In order to solve the system of equations represented by the matrix for the whole structure, boundary conditions have to be used, i.e. known values of displacements and/or forces used. Because we are dealing with structures which are in equilibrium there must be no resultant force acting on it and so we must have the sum of the forces equal to zero. Thus for the three-node system in figure 7.3,

$$F_1 + F_2 + F_3 = 0 \qquad [16]$$

Example

Figure 7.4 shows two in-line rods. Write the stiffness matrix for the structure.

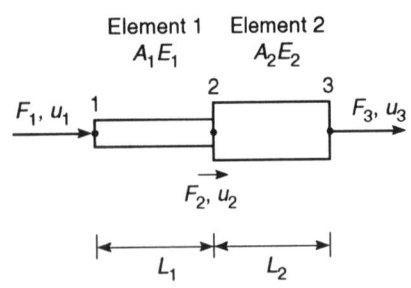

Element 1   Element 2
$A_1 E_1$ $A_2 E_2$

F_1, u_1

F_2, u_2

L_1 L_2

F_3, u_3

Fig. 7.4 Two in-line rods

The stiffness of rod 1 is $A_1 E_1 / L_1$ and that of rod 2 is $A_2 E_2 / L_2$. Thus the stiffness matrix of the structure is

$$\begin{bmatrix} \dfrac{A_1 E_1}{L_1} & -\dfrac{A_1 E_1}{L_1} & 0 \\[2ex] -\dfrac{A_1 E_1}{L_1} & \dfrac{A_1 E_1}{L_1} + \dfrac{A_2 E_2}{L_2} & -\dfrac{A_2 E_2}{L_2} \\[2ex] 0 & -\dfrac{A_2 E_2}{L_2} & \dfrac{A_2 E_2}{L_2} \end{bmatrix}$$

Example

Write the stiffness matrix for a structure represented by three springs in-line, the springs having stiffnesses of k_1, k_2 and k_3.

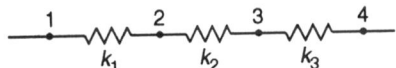

k_1 k_2 k_3

Fig. 7.5 Example

Figure 7.5 illustrates the arrangement. We have four nodes. We will thus have a 4×4 matrix.

For location $_{11}$ in the matrix we will have the sum of the stiffnesses of the elements meeting at node 1, i.e. k_1. For location $_{12}$ in the matrix we have minus the stiffness of the element joining node 1 to node 2, i.e. $-k_1$. For location $_{13}$, since there are no elements joining nodes 1 and 3 we have 0. For location $_{14}$, since there are no elements joining nodes 1 and 4 we have 0.

For location $_{21}$ in the matrix we have minus the stiffness of the element joining node 2 to node 1, i.e. $-k_1$. For location $_{22}$ we have the sum of the stiffnesses of the elements meeting at node 2, i.e. $k_1 + k_2$. For location $_{23}$ we have minus the stiffness of the element joining node 2 to node 3, i.e. $-k_2$. For location $_{24}$, since there is no element joining nodes 2 to 4 we have 0.

The entries for the other rows in the matrix can be tackled in a similar way. The result is the stiffness matrix

$$\begin{bmatrix} k_1 & -k_1 & 0 & 0 \\ -k_1 & k_1 + k_2 & -k_2 & 0 \\ 0 & -k_2 & k_2 + k_3 & -k_3 \\ 0 & 0 & -k_3 & k_3 \end{bmatrix}$$

Example

A structure has a stiffness matrix of

$$\begin{bmatrix} 100 & -100 & 0 \\ -100 & 150 & -50 \\ 0 & -50 & 50 \end{bmatrix} \text{MN/m}$$

Determine the forces acting at each node of the structure if the displacements at the nodes are $u_1 = 0.1$ mm, $u_2 = 0.2$ mm, $u_3 = 0$.

The matrix equation is

$$\begin{bmatrix} 100 & -100 & 0 \\ -100 & 150 & -50 \\ 0 & -50 & 50 \end{bmatrix} \begin{bmatrix} 0.1 \\ 0.2 \\ 0 \end{bmatrix} = \begin{bmatrix} F_1 \\ F_2 \\ F_3 \end{bmatrix}$$

Hence

$$F_1 = 100 \times 0.1 - 100 \times 0.2 + 0 = -10 \text{ kN}$$

$$F_2 = -100 \times 0.1 + 150 \times 0.2 + 0 = 20 \text{ kN}$$

$$F_3 = 0 - 50 \times 0.2 + 0 = -10 \text{ kN}$$

Note that $F_1 + F_2 + F_3 = 0$, as required by equation [16].

Review problems

2   Write the stiffness matrix for the structure shown in figure 7.6 of three rods of different materials welded together.

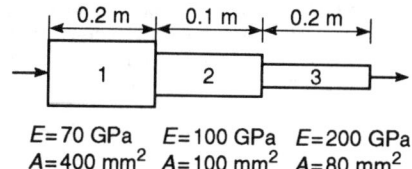

$E = 70$ GPa $E = 100$ GPa $E = 200$ GPa
$A = 400$ mm^2 $A = 100$ mm^2 $A = 80$ mm^2

Fig. 7.6 Problem 2

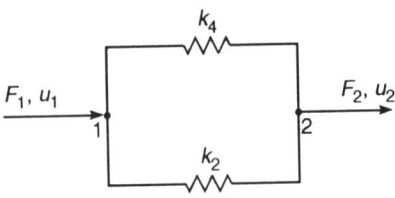

Fig. 7.7 Problem 4

3 For the structure shown in figure 7.6, the displacements of the elements are monitored and found to be $u_1 = 0$, $u_2 = 0.05$ mm, $u_3 = 0.20$ mm and $u_4 = 0.30$ mm. Hence determine the forces at each of the nodes in the structure.

4 Write the stiffness matrix for the arrangement of two springs in parallel shown in figure 7.7.

5 Write the stiffness matrix for the arrangement of springs shown in figure 7.8.

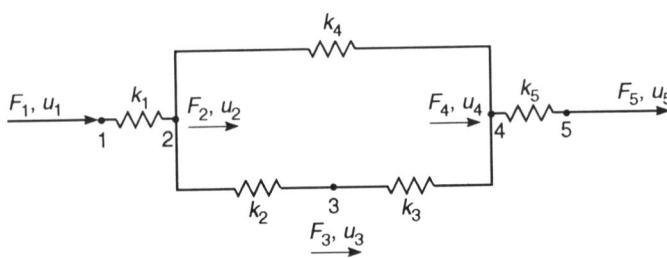

Fig. 7.8 Problem 5

7.3 Analysis of frameworks

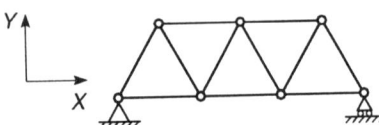

Fig. 7.9 Pin-jointed framework

Consider a framework (figure 7.9) consisting of pin-jointed members. With such types of joints there can only be tension or compression existing in the members, i.e. only forces acting along the axis of a member. We can regard each such member in the framework as being modelled by a spring. However, the spring elements are at different angles to each other, unlike the members discussed in the previous section where each spring was in-line along the x-axis. There the forces and deformations of each spring can be considered to have been defined in terms of a *local co-ordinate system*, the x-axis always being along the axis of each spring. We thus need to express the forces and deformations for each element in terms of the coordinate system of the framework as a whole. This system is generally referred to as the *global coordinate system*.

Consider a spring element, between the nodes 1 and 2, which has its local coordinate system at some angle to the global co-ordinate system, as in figure 7.10. The forces and displacements at each of the nodes can be resolved into components in the x- and y-directions. Because the element is pin-jointed, there will only be forces in the x-direction, so the F_y components will be zero. Thus we can write

$$\begin{bmatrix} k & -k \\ -k & k \end{bmatrix} \begin{bmatrix} u_{x1} \\ u_{x2} \end{bmatrix} = \begin{bmatrix} F_{x1} \\ F_{x2} \end{bmatrix}$$

We can include the F_y components, with their zero values, in this equation by rewriting it in the form

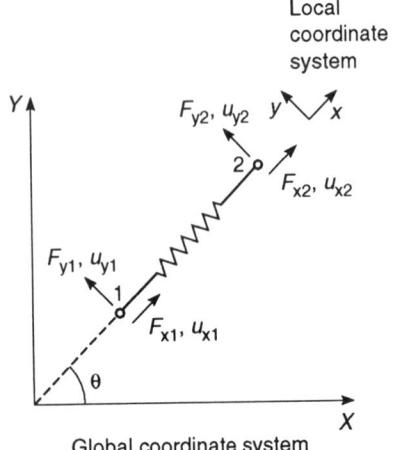

Global coordinate system

Fig. 7.10 Coordinate systems

$$
\begin{bmatrix} k & 0 & -k & 0 \\ 0 & 0 & 0 & 0 \\ -k & 0 & k & 0 \\ 0 & 0 & 0 & 0 \end{bmatrix} \begin{bmatrix} u_{x1} \\ u_{y1} \\ u_{x2} \\ u_{y2} \end{bmatrix} = \begin{bmatrix} F_{x1} \\ F_{y1} \\ F_{x1} \\ F_{y2} \end{bmatrix} \quad [17]
$$

or $f_L = K_e u_L$, where f_L is the matrix referring to forces in the local coordinate system, u_L to displacements in the local coordinate system and K_e is the element stiffness matrix.

Displacements and forces in the local coordinate directions can be derived from displacements and forces in the global coordinate directions. Thus if we have displacements at node 1 of u_{x1} and u_{y1} in the global coordinate directions we will have displacements in the local coordinate directions of

$$u_{x1} = u_{X1}\cos\theta + u_{Y1}\sin\theta$$

$$u_{y1} = -u_{X1}\sin\theta + u_{Y1}\cos\theta$$

Likewise for node 2 with displacements of u_{x2} and u_{y2} in the global coordinate directions, we have displacements in the local coordinate directions of

$$u_{x2} = u_{X2}\cos\theta + u_{Y2}\sin\theta$$

$$u_{y2} = -u_{X2}\sin\theta + u_{Y2}\cos\theta$$

These equations can be expressed in matrix form as

$$
\begin{bmatrix} u_{x1} \\ u_{y1} \\ u_{x2} \\ u_{y2} \end{bmatrix} = \begin{bmatrix} \cos\theta & \sin\theta & 0 & 0 \\ -\sin\theta & \cos\theta & 0 & 0 \\ 0 & 0 & \cos\theta & \sin\theta \\ 0 & 0 & -\sin\theta & \cos\theta \end{bmatrix} \begin{bmatrix} u_{X1} \\ u_{Y1} \\ u_{X2} \\ u_{Y2} \end{bmatrix} \quad [18]
$$

or $u_L = T u_G$, where u_L refers to the local coordinate system, u_G to the global coordinate system and T is the transformation matrix.

Similarly for the forces we can write

$$
\begin{bmatrix} F_{x1} \\ F_{y1} \\ F_{x2} \\ F_{y2} \end{bmatrix} = \begin{bmatrix} \cos\theta & \sin\theta & 0 & 0 \\ -\sin\theta & \cos\theta & 0 & 0 \\ 0 & 0 & \cos\theta & \sin\theta \\ 0 & 0 & -\sin\theta & \cos\theta \end{bmatrix} \begin{bmatrix} F_{X1} \\ F_{Y1} \\ F_{X2} \\ F_{Y2} \end{bmatrix} \quad [19]
$$

or $f_L = T f_G$, where f_L refers to the local coordinate system, f_G to the global coordinate system and T is the transformation matrix.

Thus equation [17] can be written as

$$\mathbf{f}_L = \mathbf{K}_e \mathbf{u}_L$$

$$\mathbf{T}\mathbf{f}_G = \mathbf{K}_e \mathbf{T}\mathbf{u}_G$$

Multiplying each side by the inverse of the transformation matrix, then

$$\mathbf{T}^{-1}\mathbf{T}\mathbf{f}_G = \mathbf{T}^{-1}\mathbf{K}_e \mathbf{T}\mathbf{u}_G$$

But $\mathbf{T}^{-1}\mathbf{T} = \mathbf{I}$ and so we have

$$\mathbf{f}_G = \mathbf{T}^{-1}\mathbf{K}_e \mathbf{T}\mathbf{u}_G$$

We can write this in terms of a stiffness element for the element in global coordinates $\mathbf{K}_G$ as

$$\mathbf{f}_G = \mathbf{K}_G \mathbf{u}_G \tag{20}$$

with

$$\mathbf{K}_G = \mathbf{T}^{-1}\mathbf{K}_e \mathbf{T}$$

We can find the inverse matrix by, for example, row operations (see section 3.6.2); however, it turns out that the matrix is a special type of matrix, called an *orthogonal matrix*. For such a matrix the inverse matrix is the transposed matrix (section 3.2.3). Thus we have $\mathbf{T}^{-1} = \mathbf{T}^T$. Hence

$$\mathbf{K}_G = \mathbf{T}^T \mathbf{K}_e \mathbf{T}$$

Using c to represent $\cos\theta$ and s to represent $\sin\theta$, we thus have

$$\mathbf{K}_G = \begin{bmatrix} c & -s & 0 & 0 \\ s & c & 0 & 0 \\ 0 & 0 & c & -s \\ 0 & 0 & s & c \end{bmatrix} \begin{bmatrix} k & 0 & -k & 0 \\ 0 & 0 & 0 & 0 \\ -k & 0 & k & 0 \\ 0 & 0 & 0 & 0 \end{bmatrix} \begin{bmatrix} c & s & 0 & 0 \\ -s & c & 0 & 0 \\ 0 & 0 & c & s \\ 0 & 0 & -s & c \end{bmatrix}$$

and so, after multiplying out the matrices, we can write

$$\mathbf{K}_G = k \begin{bmatrix} c^2 & cs & -c^2 & -cs \\ cs & s^2 & -cs & -s^2 \\ -c^2 & -cs & c^2 & cs \\ -cs & -s^2 & cs & s^2 \end{bmatrix} \tag{21}$$

The following example illustrates the use of the above principles in the solution of a simple framework. The framework has only a few members, to keep the calculations down to a relatively low level. Computers would be used for a more complicated structure.

Example

Determine the displacements and forces at each node for the structure shown in figure 7.11(a). The members have AE values of 200 MN.

Figure 7.11(b) shows the node diagrams for the members. The stiffness of the sloping rod is $AE/L = 200/2 = 100$ MN/m. Since we have $\sin \theta = 1/2$ and $\cos \theta = \sqrt{3}/2$ then the stiffness matrix for this rod is given by equation [21] as

$$\mathbf{K_G} = k \begin{bmatrix} c^2 & cs & -c^2 & -cs \\ cs & s^2 & -cs & -s^2 \\ -c^2 & -cs & c^2 & cs \\ -cs & -s^2 & cs & s^2 \end{bmatrix}$$

$$= 100 \begin{bmatrix} \frac{3}{2} & \frac{\sqrt{3}}{4} & -\frac{3}{2} & -\frac{\sqrt{3}}{4} \\ \frac{\sqrt{3}}{4} & \frac{1}{4} & -\frac{\sqrt{3}}{4} & -\frac{1}{4} \\ -\frac{3}{2} & -\frac{\sqrt{3}}{4} & \frac{3}{2} & \frac{\sqrt{3}}{4} \\ -\frac{\sqrt{3}}{4} & -\frac{1}{4} & \frac{\sqrt{3}}{4} & \frac{1}{4} \end{bmatrix} \text{MN/m}$$

Thus we can write, using equation [20],

$$\begin{bmatrix} F_{x1} \\ F_{y1} \\ F_{x12} \\ F_{y12} \end{bmatrix} = 100 \begin{bmatrix} \frac{3}{2} & \frac{\sqrt{3}}{4} & -\frac{3}{2} & -\frac{\sqrt{3}}{4} \\ \frac{\sqrt{3}}{4} & \frac{1}{4} & -\frac{\sqrt{3}}{4} & -\frac{1}{4} \\ -\frac{3}{2} & -\frac{\sqrt{3}}{4} & \frac{3}{2} & \frac{\sqrt{3}}{4} \\ -\frac{\sqrt{3}}{4} & -\frac{1}{4} & \frac{\sqrt{3}}{4} & \frac{1}{4} \end{bmatrix} \begin{bmatrix} u_{x1} \\ u_{y1} \\ u_{x2} \\ u_{y2} \end{bmatrix} \text{MN}$$

The stiffness of the horizontal rod is given by AE/L and is thus $200/(2 \cos 30°) = 100/\sqrt{3}$ MN/m. For this element $\cos \theta = 1$ and $\sin \theta = 0$, thus equation [21] gives

$$\mathbf{K_G} = \frac{100}{\sqrt{3}} \begin{bmatrix} 1 & 0 & -1 & 0 \\ 0 & 0 & 0 & 0 \\ -1 & 0 & 1 & 0 \\ 0 & 0 & 0 & 0 \end{bmatrix} \text{MN/m}$$

Thus we can write

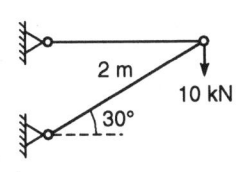

(a)

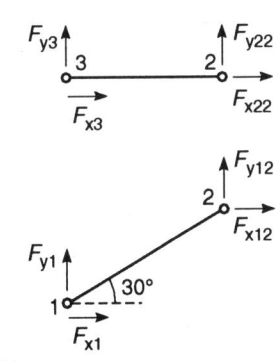

(b)

Fig. 7.11 Example

$$
\begin{bmatrix} F_{x22} \\ F_{y22} \\ F_{x3} \\ F_{y3} \end{bmatrix} = \frac{100}{\sqrt{3}} \begin{bmatrix} 1 & 0 & -1 & 0 \\ 0 & 0 & 0 & 0 \\ -1 & 0 & 1 & 0 \\ 0 & 0 & 0 & 0 \end{bmatrix} \begin{bmatrix} u_{x2} \\ u_{y2} \\ u_{x3} \\ u_{y3} \end{bmatrix} \text{MN}
$$

We can expand each of the above matrices with rows and columns of zeros so that they are the same form and we can add them. Thus they become

$$
\begin{bmatrix} F_{x1} \\ F_{y1} \\ F_{x12} \\ F_{y12} \\ 0 \\ 0 \end{bmatrix} = 100 \begin{bmatrix} \frac{3}{2} & \frac{\sqrt{3}}{4} & -\frac{3}{2} & -\frac{\sqrt{3}}{4} & 0 & 0 \\ \frac{\sqrt{3}}{4} & \frac{1}{4} & -\frac{\sqrt{3}}{4} & -\frac{1}{4} & 0 & 0 \\ -\frac{3}{2} & -\frac{\sqrt{3}}{4} & \frac{3}{2} & \frac{\sqrt{3}}{4} & 0 & 0 \\ -\frac{\sqrt{3}}{4} & -\frac{1}{4} & \frac{\sqrt{3}}{4} & \frac{1}{4} & 0 & 0 \\ 0 & 0 & 0 & 0 & 0 & 0 \\ 0 & 0 & 0 & 0 & 0 & 0 \end{bmatrix} \begin{bmatrix} u_{x1} \\ u_{y1} \\ u_{x2} \\ u_{y2} \\ u_{x3} \\ u_{y3} \end{bmatrix}
$$

$$
\begin{bmatrix} 0 \\ 0 \\ F_{x22} \\ F_{y22} \\ F_{x3} \\ F_{y3} \end{bmatrix} = \frac{100}{\sqrt{3}} \begin{bmatrix} 0 & 0 & 0 & 0 & 0 & 0 \\ 0 & 0 & 0 & 0 & 0 & 0 \\ 0 & 0 & 1 & 0 & -1 & 0 \\ 0 & 0 & 0 & 0 & 0 & 0 \\ 0 & 0 & -1 & 0 & 1 & 0 \\ 0 & 0 & 0 & 0 & 0 & 0 \end{bmatrix} \begin{bmatrix} u_{x1} \\ u_{y1} \\ u_{x2} \\ u_{y2} \\ u_{x3} \\ u_{y3} \end{bmatrix}
$$

Since $F_{x2} = F_{x12} + F_{x22}$ and $F_{y2} = F_{y12} + F_{u22}$, then

$$
\begin{bmatrix} F_{x1} \\ F_{y1} \\ F_{x2} \\ F_{y2} \\ F_{x3} \\ F_{y3} \end{bmatrix} = 100 \begin{bmatrix} \frac{3}{2} & \frac{\sqrt{3}}{4} & -\frac{3}{2} & -\frac{\sqrt{3}}{4} & 0 & 0 \\ \frac{\sqrt{3}}{4} & \frac{1}{4} & -\frac{\sqrt{3}}{4} & -\frac{1}{4} & 0 & 0 \\ -\frac{3}{2} & -\frac{\sqrt{3}}{4} & \frac{3}{2}+\frac{1}{\sqrt{3}} & \frac{\sqrt{3}}{4} & -\frac{1}{\sqrt{3}} & 0 \\ -\frac{\sqrt{3}}{4} & -\frac{1}{4} & \frac{\sqrt{3}}{4} & \frac{1}{4} & 0 & 0 \\ 0 & 0 & -\frac{1}{\sqrt{3}} & 0 & \frac{1}{\sqrt{3}} & 0 \\ 0 & 0 & 0 & 0 & 0 & 0 \end{bmatrix} \begin{bmatrix} u_{x1} \\ u_{y1} \\ u_{x2} \\ u_{y2} \\ u_{x3} \\ u_{y3} \end{bmatrix}
$$

We can now enter the boundary conditions for the structure. At the fixed supports we will have zero displacements, thus we have $u_{x1} = u_{y1} = u_{x3} = u_{y3} = 0$. At node 2 we do not know the displacements but we do know that $F_{x2} = 0$ and $F_{y2} = -10$ kN or -0.01 MN. Thus

$$
\begin{bmatrix}
F_{x1} \\
F_{y1} \\
0 \\
-0.01 \\
F_{x3} \\
F_{y3}
\end{bmatrix}
$$

$$
= 100
\begin{bmatrix}
1.5 & 0.43 & -1.5 & -0.43 & 0 & 0 \\
0.43 & 0.25 & -0.43 & -0.25 & 0 & 0 \\
-1.5 & -0.43 & 2.08 & 0.43 & -0.58 & 0 \\
-0.43 & -0.25 & 0.43 & 0.25 & 0 & 0 \\
0 & 0 & -0.58 & 0 & 0.58 & 0 \\
0 & 0 & 0 & 0 & 0 & 0
\end{bmatrix}
$$

$$
\times
\begin{bmatrix}
0 \\
0 \\
u_{x2} \\
u_{y2} \\
0 \\
0
\end{bmatrix}
\text{MN}
$$

Thus

$$
F_{x1} = 100(-1.5u_{x2} - 0.43u_{y2})
$$

$$
F_{y1} = 100(-0.43u_{x2} - 0.25u_{y2})
$$

$$
0 = 100(2.08u_{x2} - 0.43u_{y2})
$$

$$
-0.01 = 100(0.43u_{x2} + 0.25u_{y2})
$$

$$
F_{x3} = 100(-0.58u_{x2})
$$

$$
F_{y3} = 0
$$

and so we have $u_{y2} = 0.29$ mm, $u_{x2} = 0.06$ mm, $F_{x1} = -45$ kN, $F_{y1} = -9.8$ kN and $F_{x3} = -3.5$ kN.

Review problems

6   Determine the displacements and forces at each node for the structure shown in figure 7.12. The members have AE values of 200 MN.

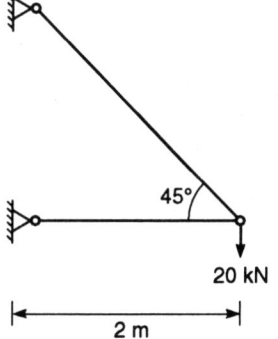

45°

20 kN

2 m

Fig. 7.12 Problem 6

7.4 The finite element method

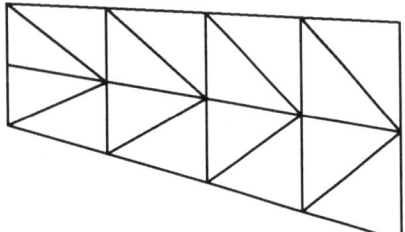

Fig. 7.13 Model of a two-dimensional shape

The above consideration of a framework serves to introduce the idea of a structure being represented by an array of elements, linear spring elements, joined together at their node points. We can also use a similar model to represent two-dimensional or three-dimensional solids. Figure 7.13 shows how we might model a two-dimensional shape, the continuous solid being modelled by an array of plane elements which are joined together at the node points. The choice of element shape and size is dictated by the geometry of the body and the accuracy required, bearing in mind that the more elements there are the greater the number of computations. The distribution of forces and displacements within the solid can then be determined in a similar way to that used for the frameworks. This method is termed the *finite element method*. The term 'finite element' occurs because the infinity of points within a solid is replaced by a number of finite points, the nodes.

The method is used for a wide diversity of problems, e.g. the elastic behaviour of two- and three-dimensional solids, heat conduction in two- and three-dimensional solids, fluid dynamics, diffusion, electric and magnetic fields.

Further problems

7 Write the stiffness matrices for the following arrangements of springs:
   (a) two springs, stiffnesses k_1 and k_2 in-line,
   (b) three springs, stiffnesses k_1, k_2 and k_3 in-line,
   (c) three springs, each of stiffness k, in-line,
   (d) four springs, each of stiffness k in-line.
8 Determine the displacements at the loaded node for each of the structures shown in figure 7.14. The members have AE values of 200 MN.

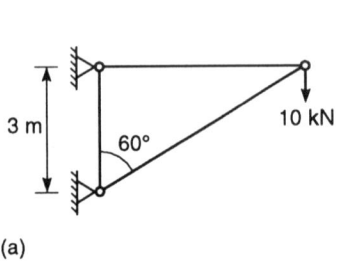

(a)

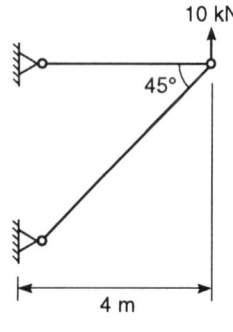

(b)

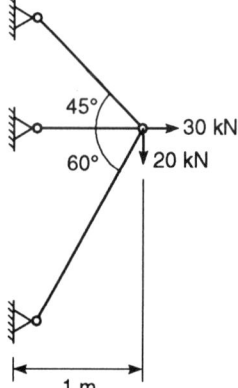

(c)

Fig. 7.14 Problem 8

8 Iterative methods

8.1 Iteration techniques

The techniques for the solution of simultaneous equations that have been discussed so far in this book, e.g. Gaussian elimination, can be termed *direct methods*. However there is a major drawback to direct methods; the number of computations involved with a large set of equations is high. An alternative group of techniques are *iterative methods*. These generate a sequence of approximate solutions which may converge to the required solution. They have the advantage of being easy to program for computer solution. They also have the advantage, when being worked out manually, of being self-correcting if an error is made. Iterative techniques can be crudely put as: to start off guess a solution, use it to obtain an improved solution, use the improved solution to obtain a yet further improved solution, and keep on repeating the process until a solution of sufficient accuracy is obtained. In this chapter Jacobi and Gauss–Seidel iteration techniques are discussed.

8.2 Jacobi iteration

Consider a pair of simultaneous equations,

$$2x + y = 4$$

$$x + 3y = 7$$

We first rewrite the equations in the form

$$x = -\tfrac{1}{2}y + 2 \qquad [1]$$

$$y = -\tfrac{1}{3}x + \tfrac{7}{3} \qquad [2]$$

We now guess a solution. Suppose we guess $x = 1$, $y = 1$. We now substitute these values in the right-hand sides of equations [1] and [2] and obtain new estimates. Thus equation [1] now gives

$$x = -\tfrac{1}{2}y + 2 = -0.5 + 2 = 1.5$$

and equation [2]

$$y = -\tfrac{1}{3}x + \tfrac{7}{3} = -\tfrac{1}{3} + \tfrac{7}{3} = 2.0$$

We now take these estimates and again substitute them in the right-hand sides of equations [1] and [2]. Thus equation [1] now gives

$$x = -\tfrac{1}{2}y + 2 = -1.0 + 2 = 1.0$$

and equation [2]

$$y = -\tfrac{1}{3}x + \tfrac{7}{3} = -\tfrac{1.5}{3} + \tfrac{7}{3} = 1.83$$

We now take these estimates and again substitute them in the right-hand sides of equations [1] and [2]. Thus equation [1] now gives

$$x = -\tfrac{1}{2}y + 2 = -0.92 + 2 = 1.08$$

and equation [2] gives

$$y = -\tfrac{1}{3}x + \tfrac{7}{3} = -\tfrac{1}{3} + \tfrac{7}{3} = 2.0$$

We now take these estimates and again substitute them in the right-hand sides of equations [1] and [2]. Thus equation [1] now gives

$$x = -\tfrac{1}{2}y + 2 = -1.0 + 2 = 1.0$$

and equation [2] gives

$$y = -\tfrac{1}{3}x + \tfrac{7}{3} = -\tfrac{1.08}{3} + \tfrac{7}{3} = 1.97$$

The values are converging to $x = 1$ and $y = 2$.

The procedure adopted for a set of simultaneous equations with the variable x_1, x_2, x_3, etc. can be summarised as:

1 Rearrange the equations so that you have an equation for each variable in the form $x_1 = ...$, $x_2 = ...$, etc. Wherever possible, use the equation with the largest coefficient of the variable. For example, if we have $x + 4y = 3$, then preferentially select this equation for the y variable rather than the x variable.
2 Make an initial guess as to the values of each variable.
3 Substitute the values in the right-hand side of the equations

given in step 1 and obtain new values of the variables.
4   Repeat step 3 until successive values of the variables are sufficiently alike.

Example

Use the Jacobi iteration technique to solve the following set of simultaneous equations:

$$4x + 2y + z = 9, \ x + y + 2z = 1, \ x + 3y - z = 6$$

We will use the first equation for the x variable, with the second equation for the z variable and the third equation for the y variable. This enables us to select the equations with the variable with the largest coefficient. Rewriting the equations gives

$$x = -\tfrac{1}{2}y - \tfrac{1}{4}z + \tfrac{9}{4} \qquad\qquad [3]$$

$$y = -\tfrac{1}{3}x + \tfrac{1}{3}z + 2 \qquad\qquad [4]$$

$$z = -\tfrac{1}{2}x - \tfrac{1}{2}y + \tfrac{1}{2} \qquad\qquad [5]$$

For an initial guess we will take $x = 0$, $y = 0$ and $z = 0$. Using these values in equations [3], [4] and [5] gives improved values $x = 9/4$, $y = 2$ and $z = 1/2$. Taking these values and using them in equations [3], [4] and [5] gives

$$x = -\tfrac{1}{2}y - \tfrac{1}{4}z + \tfrac{9}{4} = -1 - 0.125 + 2.25 = 1.125$$

$$y = -\tfrac{1}{3}x + \tfrac{1}{3}z + 2 = -0.75 + 0.167 + 2 = 1.417$$

$$z = -\tfrac{1}{2}x - \tfrac{1}{2}y + \tfrac{1}{2} = -1.125 - 1 + 0.5 = -1.625$$

Taking these values and using them in equations [3], [4] and [5] gives

$$x = -\tfrac{1}{2}y - \tfrac{1}{4}z + \tfrac{9}{4} = -0.709 + 0.406 + 2.25 = 1.947$$

$$y = -\tfrac{1}{3}x + \tfrac{1}{3}z + 2 = -0.375 - 0.542 = 1.083$$

$$z = -\tfrac{1}{2}x - \tfrac{1}{2}y + \tfrac{1}{2} = -0.563 - 0.709 + 0.5 = -0.772$$

Taking these values and using them in equations [3], [4] and [5] gives

$$x = -\tfrac{1}{2}y - \tfrac{1}{4}z + \tfrac{9}{4} = -0.542 + 0.193 + 2.25 = 1.901$$

$$y = -\tfrac{1}{3}x + \tfrac{1}{3}z + 2 = -0.649 - 0.257 + 2 = 1.094$$

$$z = -\tfrac{1}{2}x - \tfrac{1}{2}y + \tfrac{1}{2} = -0.974 - 0.542 + 0.5 = -1.016$$

Taking these values and using them in equations [3], [4] and [5] gives

$$x = -\tfrac{1}{2}y - \tfrac{1}{4}z + \tfrac{9}{4} = -0.547 + 0.254 + 2.25 = 1.957$$

$$y = -\tfrac{1}{3}x + \tfrac{1}{3}z + 2 = -0.634 - 0.339 + 2 = 1.028$$

$$z = -\tfrac{1}{2}x - \tfrac{1}{2}y + \tfrac{1}{2} = -0.951 - 0.547 + 0.5 = -0.998$$

Taking these values and using them in equations [3], [4] and [5] gives

$$x = -\tfrac{1}{2}y - \tfrac{1}{4}z + \tfrac{9}{4} = -0.514 + 0.250 + 2.25 = 1.986$$

$$y = -\tfrac{1}{3}x + \tfrac{1}{3}z + 2 = -0.652 - 0.333 + 2 = 1.015$$

$$z = -\tfrac{1}{2}x - \tfrac{1}{2}y + \tfrac{1}{2} = -0.979 - 0.514 + 0.5 = -0.993$$

The results so far are:

	1st	2nd	3rd	4th	5th	6th
x	0	1.125	1.947	1.901	1.958	1.986
y	0	1.417	1.083	1.094	1.028	1.015
z	0	1.625	-0.772	-1.016	-0.998	-0.993

To an accuracy of one decimal place, the results are converging to $x = 2.0$, $y = 1.0$ and $z = -1.0$.

Review problems

1   Use the Jacobi iteration technique to solve the following sets of simultaneous equations:

(a) $2x + y = 1$, $x - 3y = 5$,

(b) $x + 2y = 7$, $4x - y = 1$,

(c) $4x + y + 2z = 9$, $x + y + 3z = 7$, $x + 3y - z = 9$,

(d) $4x - 2y + z = -9$, $2x + 4y + z = 1$, $x + 2y + 4z = 4$,

(e) $x + y + 4z = 12$, $x + 3y + z = 8$, $3x + y - z = 8$,

(f) $4x + 2y - z = 4$, $x + 3y + z = 6$, $x + 2y - 5z = -7$

8.3 Gauss–Seidel iteration

With the Jacobi iteration technique we can calculate a set of values of the variables and then use them to obtain the next set. However, after the first equation has been solved we have a better value for the first variable. Rather than use the older value of this variable in the second equation, we can use the newer value. This variation of the method, with the newest version of variables being used, is called the *Gauss–Seidel iteration technique*.

The technique can be summarised as:

1 Rearrange the equations so that you have an equation for each variable in the form $x_1 = ...$, $x_2 = ...$, etc. Wherever possible, use the equation with the largest coefficient of the variable.
2 Make an initial guess as to the values of each variable.
3 Substitute the values in the right-hand side of the equation for the first variable given in step 1 and obtain a new value of the first variable.
4 Use this newer value of the first variable and the older values of the other variables to obtain a new value for the second variable.
5 Keep on using the latest values of the variables to obtain new values of each variable.
6 Repeat step 5 until successive values of the variables are sufficiently alike.

The following illustrates the technique with the two equations [1] and [2] used in the previous section to illustrate the Jacobi method.

$$x = -\tfrac{1}{2}y + 2 \qquad\qquad [1]$$

$$y = -\tfrac{1}{3}x + \tfrac{7}{3} \qquad\qquad [2]$$

We now guess a solution. Suppose we guess $x = 1$, $y = 1$. We now substitute these values in the right-hand sides of equations [1] and [2] and obtain new estimates. Thus equation [1] now gives

$$x = -\tfrac{1}{2}y + 2 = -0.6 + 2 = 1.5$$

Using this newer estimate of x with equation [2] gives

$$y = -\tfrac{1}{3}x + \tfrac{7}{3} = -0.5 + 2.33 = 1.83$$

Using this newer estimate of y with equation [1] gives

$$x = -\tfrac{1}{2}y + 2 = -0.92 + 2 = 1.08$$

Using this newer estimate of x with equation [2] gives

$$y = -\tfrac{1}{3}x + \tfrac{7}{3} = -0.36 + 2.33 = 1.97$$

Using this newer estimate of y with equation [1] gives

$$x = -\tfrac{1}{2}y + 2 = -0.99 + 2 = 1.01$$

Using this newer estimate of x with equation [2] gives

$$y = -\tfrac{1}{3}x + \tfrac{7}{3} = -0.34 + 2.33 = 1.99$$

The results have more rapidly converged to the values of $x = 1$ and $y = 2$. The Gauss–Seidel technique is usually better than the Jacobi iteration in that fewer steps are needed to obtain a particular accuracy.

Example

Use the Gauss–Seidel iteration technique to solve the following set of simultaneous equations:

$$4x + 2y + z = 9, \quad x + y + 2z = 1, \quad x + 3y - z = 6$$

This is the same example as used in the previous section with the Jacobi technique. We will use the first equation for the x variable, the second equation for the z variable and the third equation for the y variable. This enables us to select the equations with the variable with the largest coefficient. Rewriting the equations gives

$$x = -\tfrac{1}{2}y - \tfrac{1}{4}z + \tfrac{9}{4} \qquad\qquad [3]$$

$$y = -\tfrac{1}{3}x + \tfrac{1}{3}z + 2 \qquad\qquad [4]$$

$$z = -\tfrac{1}{2}x - \tfrac{1}{2}y + \tfrac{1}{2} \qquad\qquad [5]$$

For an initial guess we will take $y = 0$ and $z = 0$. Using these values in equation [3] gives

$$x = -\tfrac{1}{2}y - \tfrac{1}{4}z + \tfrac{9}{4} = 0 - 0 + 2.25 = 2.25$$

Using $x = 2.25$, and $z = 0$ in equation [4] gives

$$y = -\tfrac{1}{3}x + \tfrac{1}{3}z + 2 = -0.75 + 0 + 2 = 1.25$$

Using $x = 2.25$ and $y = 1.25$ in equation [5] gives

$$z = -\tfrac{1}{2}x - \tfrac{1}{2}y + \tfrac{1}{2} = -1.125 - 0.625 + 0.5 = -1.25$$

Using $y = 1.25$ and $z = -1.25$ in equation [3] gives

$$x = -\tfrac{1}{2}y - \tfrac{1}{4}z + \tfrac{9}{4} = -0.625 + 0.313 + 2.25 = 1.938$$

Using $x = 1.938$ and $z = -1.25$ in equation [4] gives

$$y = -\tfrac{1}{3}x + \tfrac{1}{3}z + 2 = -0.646 - 0.417 + 2 = 0.937$$

Using $x = 1.938$ and $y = 0.937$ in equation [5] gives

$$z = -\tfrac{1}{2}x - \tfrac{1}{2}y + \tfrac{1}{2} = -0.969 - 0.469 + 0.5 = -0.938$$

Using $y = 0.937$ and $z = -0.938$ in equation [3] gives

$$x = -\tfrac{1}{2}y - \tfrac{1}{4}z + \tfrac{9}{4} = -0.469 + 0.235 + 2.25 = 2.016$$

Using $x = 2.016$ and $z = -0.938$ in equation [4] gives

$$y = -\tfrac{1}{3}x + \tfrac{1}{3}z + 2 = -0.672 - 0.313 + 2 = 1.015$$

Using $x = 2.016$ and $y = 1.015$ in equation [5] gives

$$z = -\tfrac{1}{2}x - \tfrac{1}{2}y + \tfrac{1}{2} = -1.008 - 0.5075 + 0.5 = -1.016$$

The following are the results so far obtained at each iteration and show how they are converging.

	1st	2nd	3rd	4th
x		2.25	1.938	2.016
y	0	1.25	0.937	1.015
z	0	-1.25	-0.938	-1.016

To an accuracy of one decimal place, the results are converging to $x = 2.0$, $y = 1.0$ and $z = -1.0$. Compare the rate of convergence in this example with the example used for the Jacobi iteration; it is must faster at converging.

Review problems

2   Use the Gauss–Seidel technique with the equations given in review problem 1.

8.4 Convergence

The Jacobi and the Gauss–Seidel techniques do not always give results which converge, regardless of the number of iterations done. As a part of the sequence of instructions for carrying out the Jacobi and Gauss–Seidel techniques, the statement is included: 'Wherever possible, use the equation with the largest coefficient of the variable.' Thus for the set of equations

$$a_{11}x_1 + a_{12}x_2 + a_{13}x_3 + \ldots = c_1$$

$$a_{21}x_1 + a_{22}x_2 + a_{23}x_3 + \ldots = c_2$$

etc.

we would pick for the equation for the variable x_1 the one in which x_1 has the largest coefficient, i.e. in the first equation a_{11} is greater than the other coefficients.

This is really just a crude attempt to make the diagonal entry in the matrix of the coefficients larger in magnitude than the sum of the magnitudes of the other coefficients in that row. Such a system is said to be *diagonally dominant* and will always converge. What we have is then, for the first equation,

$$a_{11}x_1 = c_1 - a_{12}x_2 - a_{13}x_3 - \ldots$$

$$x_1 = \frac{c_1}{a_{11}} - \frac{a_{12}}{a_{11}}x_2 - \frac{a_{13}}{a_{11}}x_3 - \ldots$$

The error in the next calculated value of x_1 will be the sum of the errors in all the other x terms multiplied by their coefficients. Thus, if all the coefficients in the above equation are less than 1 then the error will decrease as the iteration proceeds.

The above condition of diagonal dominance is a guarantee that converge will occur. However if there is not diagonal dominance it does not mean that convergence will not occur. The speed with which the iterations converge depends on the dominance of the diagonal terms. The more dominant the diagonal terms the more rapidly the iterations converge.

Example

What arrangement of the following equations will improve the chance and speed of convergence?

$$4x + y - 7z = 2, \quad 7x - 2y + 2z = 1, \quad 5x + 11y - 4z = 3$$

The matrix of the coefficients is

$$\begin{bmatrix} 4 & 1 & -7 \\ 7 & -2 & 2 \\ 5 & 11 & -4 \end{bmatrix}$$

The diagonal terms are 4, −2 and −4. In the first row 4 is not greater than $|1| + |-7|$. In the second row $|-2|$ is not greater than $|7| + |2|$. In the third row $|-4|$ is not greater than $|5| + |11|$. Thus the matrix is not diagonally dominant. We can, however, rearrange the rows to give

$$\begin{bmatrix} 7 & -2 & 2 \\ 5 & 11 & -4 \\ 4 & 1 & 7 \end{bmatrix}$$

$|7|$ is greater than $|-2| + |2|$, $|11|$ is greater than the sum of $|5| + |-4|$ and $|7|$ is greater than $|4| + |1|$. Now the matrix is diagonally dominant. Thus for the variable x we choose to rearrange the equation $7x - 2y + 2z = 1$, for y the equation $5x + 11y - 4z = 3$ and for z the equation $4x + y - 7z = 2$.

Review problems

3   Which of the following are diagonally dominant?

(a) $\begin{bmatrix} 1 & -2 \\ 4 & 1 \end{bmatrix}$, (b) $\begin{bmatrix} -3 & 2 \\ 2 & -4 \end{bmatrix}$, (c) $\begin{bmatrix} 2 & -1 & 4 \\ -4 & 1 & 2 \\ 2 & -5 & 1 \end{bmatrix}$,

(d) $\begin{bmatrix} 4 & 2 & -1 \\ -2 & 5 & 2 \\ 1 & 3 & 6 \end{bmatrix}$, (e) $\begin{bmatrix} 5 & 0 & 1 & -2 \\ 3 & -8 & 2 & 1 \\ 0 & 3 & 5 & -1 \\ 2 & -1 & 2 & 6 \end{bmatrix}$

4   For those matrices in the previous problem that are not diagonally dominant, rearrange the rows to make them diagonally dominant.

Further problems

5   Use (a) the Jacobi iteration technique, (b) the Gauss–Seidel iteration technique, to solve the following set of simultaneous equations:

(a) $2x + 4y - z = 11$, $x + 2y - 4z = 2$, $3x + y + z = 9$,

(b) $3x - y + z = 6$, $x + 2y + z = 1$, $x + y + 3z = 6$,

(c) $3x + y - z = 3.9$, $x - 5y + 2z = 2.1$, $x + y + 3z = 4.1$,

(d) $x + 2y + 4z = 9$, $3x + y - z = 3$, $x + 4y + z = 6.3$,

(e) $5x + 2y + z = 1.9$, $x + 3y - z = 1.8$, $x + y - 3z = 1$

6   Using either mesh or nodal analysis and either the Jacobi or
Gauss–Seidel iteration technique, determine the currents in the
circuits given in figure 8.1.

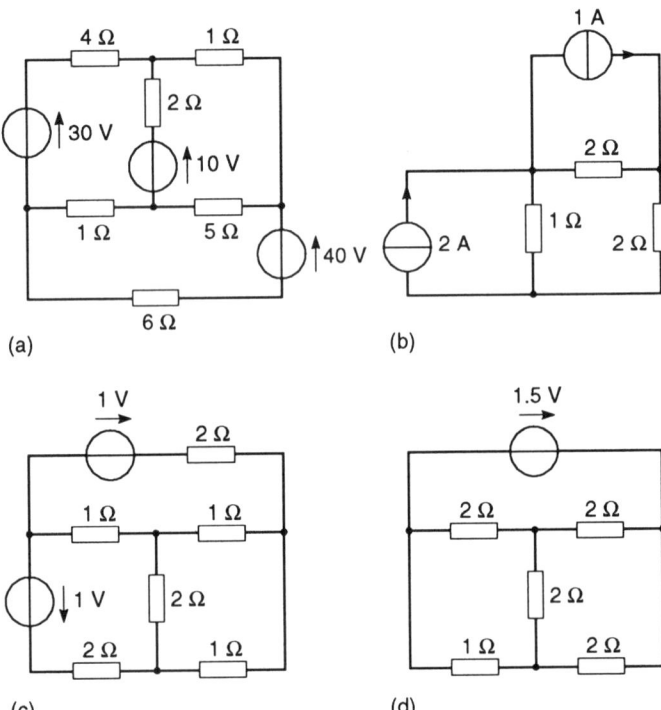

Fig. 8.1 Problem 6

Answers

1 Linear (b), (d), (e), non-linear (a), (c), (f)

2 (a) and (b) are linear. The cos 30° and cos 60° are not cosines of a variable.

3 (a) and (b) are linear

4 (a) One solution, (b) no solutions, (c) infinite number of solutions

5 (a) $x = -13/5$, $y = -11/5$, (b) $x = -1$, $y = 2$, (c) $x = -3$, $y = 2$, (d) $x = 1/5$, $y = 8/5$, (e) $x = -2$, $y = 0$, (f) $x = 2$, $y = 3$

6 $I_1 = 0.182$ A, $I_2 = -0.091$ A

7 $a = 1$, $b = 0.5$

8 $\sigma_c = 6.12$ MPa, $\sigma_s = 61.2$ MPa

9 (a) $x = 3$, $y = 4$, (b) $x = 3$, $y = -1$, (c) $x = 3$, $y = 4$

10 $x = 4$, $y = -2$, $z = 3$

11 $x = -1$, $y = 2$, $z = 1$

12 5.6 A, 2.0 A, −0.80 A

13 (a) (g) (i) (j)

14 (b) (d)

15 (a) $y = 2$, $x = 3$, (b) $y = -1$, $x = 2$, (c) $y = 2$, $x = 3$, (d) $y = 3$, $x = 2$, (e) $y = 4$, $x = 2$, (f) $y = 12$, $x = -4$

16 $I_1 = 7$ A, $I_2 = -2$ A

17 $F = 47.5$ N, (b) $T = 64.5$ N

18 $I_1 = 0.2$, $I_2 = 0.5$

19 $I_1 = 0.663$, $I_2 = 1.362$, $I_3 = 0.826$

Chapter 2

1 $x = 5, y = -1$

2 $x = 2, y = -1, z = 1$

3 (a) $\begin{bmatrix} 1 & 3 & 5 & 2 \\ 2 & -1 & 1 & 4 \\ 3 & 2 & -1 & 3 \end{bmatrix}$, (b) $\begin{bmatrix} 1 & 1 & 2 & 1 \\ -1 & 3 & 1 & 2 \\ 2 & 1 & -1 & 5 \end{bmatrix}$

4 $\begin{bmatrix} 2 & 1 & 3 & -1 & 2 \\ -1 & 3 & 1 & 2 & 1 \\ 3 & -1 & 1 & 3 & 4 \\ 1 & 2 & -3 & -1 & 1 \end{bmatrix}$

5 $2x + y + 3z = 6, -1x + 5z = 2, 4x - 2y - 1z = 1$

6 (a) $x = -5, y = 1, z = 3$, (b) $x = 2, y = 1, z = -3$,
   (c) $x = 3, y = 2, z = 1$, (d) $x = 5, y = -1, z = 2$

7 $a = -1, b = 1, c = 2, d = 1$

8 $x = 1.95, y = 1.24$, residuals 0, 0

9 $x = 0.01, y = 2.01$, residuals 0, 0.02

10 $x = 11, y = -10$ changes to $x = 6, y = -5$

11 $x = 1.00, y = 1.00, z = 1.00$

12 $x = -2, y = 3, z = 6$

13 (a) Unique, $x = 1, y = 1$, (b) infinite, (c) no solution

14 (a) $x = 1 + t, y = t - 4, z = t$, (b) inconsistent,
    (c) $x = 10 - 3t, y = 2t - 4, z = t$

15 (a) $x = 5, y = 2, z = -3$, (b) $x = 2.3, y = 3.7, z = -1.9$,
    (c) $x = 2, y = 3, z = 1$, (d) $a = 2, b = 1, c = 1, d = 3$,
    (e) $x = 2, y = 1, z = -1$, (f) $x = 1, y = 1, z = 1$

16 $F_1 = 2$ N, $F_2 = 1$ N, $F_3 = 3$ N

17 $I_1 = 2$ A, $I_2 = 1$ A, $I_3 = 1$ A

18 $i_1 = \dfrac{Z_2(e_1 - e_3) + Z_3(e_1 - e_2)}{Z_2Z_3 + Z_3Z_1 + Z_1Z_2}$, $i_2 = \dfrac{Z_1(e_2 - e_3) + Z_3(e_2 - e_1)}{Z_2Z_3 + Z_3Z_1 + Z_1Z_2}$,

    $i_3 = \dfrac{Z_1(e_3 - e_2) + Z_2(e_3 - e_1)}{Z_2Z_3 + Z_3Z_1 + Z_1Z_2}$

19 (a) infinite, $x = t, y = 2 - t, z = t$, (b) inconsistent,
    (c) infinite, $x = 2 + t, y = t - 1, z = t$,

Chapter 3

1 (a) $\begin{bmatrix} 2 & 3 & 1 \\ 1 & -2 & 2 \\ 3 & 1 & 0 \end{bmatrix}$, (b) $\begin{bmatrix} 3 & 2 \\ 1 & 1 \end{bmatrix}$, (c) $\begin{bmatrix} 1 & 2 & 1 \\ 0 & 1 & 5 \\ 1 & -2 & 1 \end{bmatrix}$

2 (a) 2×2, square, (b) 3×1, column, (c) 1×2, row,
   (d) 2×2, square, unit matrix, (e) 3×3, square, diagonal,
   (f) 3×3, square, null, (g) 3×2

3 (a) $\begin{bmatrix} 1 & 2 \\ 4 & 0 \end{bmatrix}$, (b) $\begin{bmatrix} 1 & 3 & -1 \\ 5 & 0 & 2 \end{bmatrix}$, (c) $\begin{bmatrix} 1 & 4 \\ 2 & 2 \\ 5 & 1 \end{bmatrix}$

4 **A, B**

5 (a) $\begin{bmatrix} 3 & -1 \\ 3 & 5 \end{bmatrix}$, (b) $\begin{bmatrix} 1 & 5 \\ 3 & 3 \end{bmatrix}$, (c) not possible,

(d) $\begin{bmatrix} 0 & 2 & 4 \\ 6 & 2 & 6 \end{bmatrix}$, (e) $\begin{bmatrix} -2 & -2 & -2 \\ -2 & -8 & -6 \end{bmatrix}$, (f) not possible,

(g) $\begin{bmatrix} 1 & -5 \\ -3 & -3 \end{bmatrix}$, (h) as (a).

6 (a) $\begin{bmatrix} 1000 & 600 & 800 & 1200 \\ 400 & 300 & 400 & 500 \\ 100 & 100 & 120 & 120 \\ 1200 & 1200 & 1300 & 1500 \end{bmatrix}$,

$\begin{bmatrix} 1200 & 1000 & 800 & 800 \\ 500 & 600 & 400 & 500 \\ 120 & 120 & 110 & 130 \\ 1500 & 1700 & 1800 & 1600 \end{bmatrix}$,

(b) $\begin{bmatrix} 2200 & 1600 & 1600 & 2000 \\ 900 & 900 & 800 & 1000 \\ 220 & 220 & 230 & 250 \\ 2700 & 2900 & 3100 & 3100 \end{bmatrix}$,

(c) $\begin{bmatrix} 200 & 400 & 0 & -400 \\ 100 & 300 & 0 & 0 \\ 20 & 20 & -10 & 10 \\ 300 & 500 & 500 & 100 \end{bmatrix}$

7 As given in the problem

8 (a) $\begin{bmatrix} 2 & 4 \\ 4 & 6 \\ 0 & 8 \end{bmatrix}$, (b) $\begin{bmatrix} 3 & 6 \\ 6 & 9 \\ 0 & 12 \end{bmatrix}$, (c) $\begin{bmatrix} -1 & -2 \\ -2 & -3 \\ 0 & -4 \end{bmatrix}$,

(d) $\begin{bmatrix} 0.5 & 1 \\ 1 & 1.5 \\ 0 & 2 \end{bmatrix}$

9 $\begin{bmatrix} 144 & 72 & 18 \\ 96 & 36 & 6 \end{bmatrix}$

10 (a) $\begin{bmatrix} 4 & 7 \\ 1 & 8 \\ 4 & 2 \end{bmatrix}$, (b) $\begin{bmatrix} 7 & 11 \\ 3 & 9 \\ -3 & 1 \end{bmatrix}$, (c) $\begin{bmatrix} -3 & -4 \\ -2 & -1 \\ 7 & 1 \end{bmatrix}$,

(d) $\begin{bmatrix} 0 & -1 \\ 1 & -4 \\ -8 & -2 \end{bmatrix}$, (e) $\begin{bmatrix} 5 & 8 \\ 2 & 7 \\ -1 & 1 \end{bmatrix}$

11 (a) $2\begin{bmatrix} 2 & 3 \\ 1 & 5 \end{bmatrix}$, (b) $3\begin{bmatrix} 1 & -1 & 2 \\ 3 & 2 & 4 \end{bmatrix}$, (c) $b\begin{bmatrix} a & b & b \\ a-2 & 3 & b^2 \\ ab & 1 & 1 \end{bmatrix}$

12 (a) $\begin{bmatrix} 10 & 11 \\ 8 & 14 \end{bmatrix}$, $\begin{bmatrix} 1 & 1 & 7 \\ -2 & 0 & -4 \\ -1 & 5 & 23 \end{bmatrix}$, $\begin{bmatrix} 7 \\ 10 \end{bmatrix}$, not possible,

(b) $\begin{bmatrix} 14 & 36 & 25 \\ 4 & -1 & 7 \\ 12 & 26 & 21 \end{bmatrix}$, $\begin{bmatrix} 9 & 8 & 19 \\ -2 & 0 & 0 \\ 32 & 9 & 25 \end{bmatrix}$, $\begin{bmatrix} 16 \\ 1 \\ 12 \end{bmatrix}$, $\begin{bmatrix} 9 \\ -1 \\ 12 \end{bmatrix}$,

(c) $\begin{bmatrix} 2 & -1 \\ 8 & 13 \\ 6 & 14 \end{bmatrix}$, not possible, not possible, not possible

13 $\begin{bmatrix} 3 \\ -1 \\ 2 \end{bmatrix}$

14 $\begin{bmatrix} x+4y+z \\ 2x-y+3z \end{bmatrix}$

15 $\begin{bmatrix} 2 & 4 & 9 \\ 1 & 3 & 5 \\ 2 & 5 & 8 \end{bmatrix}\begin{bmatrix} 6 \\ 12 \\ 5 \end{bmatrix} = \begin{bmatrix} 105 \\ 67 \\ 112 \end{bmatrix}$

16 $\mathbf{A}^{-1} = \frac{1}{2}\begin{bmatrix} 1 & 1 \\ -1 & 1 \end{bmatrix}$, $\mathbf{B}^{-1} = -\frac{1}{7}\begin{bmatrix} 5 & -4 \\ -2 & 3 \end{bmatrix}$,

$\mathbf{C}^{-1} = -\frac{1}{2}\begin{bmatrix} 5 & 3 \\ 4 & 2 \end{bmatrix}$, $\mathbf{D}^{-1} = -\frac{1}{10}\begin{bmatrix} 2 & -2 \\ -4 & -1 \end{bmatrix}$

17 $\mathbf{A}^{-1} = \frac{1}{6}\begin{bmatrix} 4 & -4 & 2 \\ -2 & 2 & 2 \\ 1 & 2 & -1 \end{bmatrix}$, $\mathbf{B}^{-1} = \frac{1}{7}\begin{bmatrix} 1 & 2 & 6 \\ -3 & 1 & 3 \\ 1 & 2 & -1 \end{bmatrix}$,

$\mathbf{C}^{-1} = \frac{1}{6}\begin{bmatrix} 1 & 5 & -4 \\ 1 & -1 & 2 \\ -1 & 7 & -2 \end{bmatrix}$, $\mathbf{D}^{-1} = \frac{1}{2}\begin{bmatrix} 1 & 0 & -1 \\ 1 & -1 & 2 \\ -1 & 1 & -1 \end{bmatrix}$,

$$\mathbf{E}^{-1} = \begin{bmatrix} 1 & 0 & -1 \\ 1 & -1 & 2 \\ -1 & 1 & -1 \end{bmatrix}, \mathbf{F}^{-1} = \frac{1}{9}\begin{bmatrix} -2 & 5 & -1 \\ 4 & -1 & 2 \\ -3 & 3 & 3 \end{bmatrix}$$

18 (a) $x = 0$, $y = 1$, (b) $x = 41/29$, $y = -13/29$,
    (c) $x = 11/3$, $y = 1/3$

19 (a) $x = 2$, $y = 3$, $z = 1$, (b) $x = -1$, $y = 1$, $z = 0$,
    (c) $x = 16/7$, $y = -6/7$, $z = 30/7$, (d) $x = 2$, $y = 2$, $z = 1$

20 (a) $\begin{bmatrix} 3 & 2 & 1 \\ 3 & 2 & 3 \\ 1 & 2 & 2 \end{bmatrix}$, (b) $\begin{bmatrix} -1 & 2 & 5 \\ 3 & 2 & 8 \\ -1 & -3 & 2 \end{bmatrix}$, (c) $\begin{bmatrix} 1 & 0 & -1 \\ -1 & 2 & -1 \\ 3 & 2 & -2 \end{bmatrix}$,

(d) $\begin{bmatrix} 2 & -1 & -2 \\ 0 & -1 & -2 \\ 0 & 2 & 1 \end{bmatrix}$, (e) $\begin{bmatrix} 3 & 2 & -1 \\ 2 & 1 & -1 \\ -1 & 3 & 1 \end{bmatrix}$, (f) $\begin{bmatrix} -3 & -2 & 1 \\ -2 & -1 & 1 \\ 1 & 3 & -1 \end{bmatrix}$,

(g) $\begin{bmatrix} 4 & 2 & 0 \\ 2 & 4 & 2 \\ 4 & 4 & 0 \end{bmatrix}$, (h) $\begin{bmatrix} 3 & 3 & 3 \\ 6 & 0 & 6 \\ -3 & 0 & 6 \end{bmatrix}$, (i) $\begin{bmatrix} -2 & 4 & 8 \\ 5 & 3 & 12 \\ 0 & -5 & 3 \end{bmatrix}$,

(j) $\begin{bmatrix} 0 & -1 & -2 \\ -3 & 2 & -2 \\ 4 & 2 & -4 \end{bmatrix}$, (k) $\begin{bmatrix} -4 & -2 & 2 \\ -1 & -3 & 2 \\ -2 & -5 & 1 \end{bmatrix}$,

(l) $\begin{bmatrix} 2 & 1 & 1 \\ 1.5 & 2.5 & 3 \\ 3 & 1.5 & 0.5 \end{bmatrix}$

21 (a) $\begin{bmatrix} 1 & 11 \\ -1 & 21 \end{bmatrix}$, (b) $\begin{bmatrix} 18 & 20 \\ 2 & 4 \end{bmatrix}$, (c) not possible,

(d) $\begin{bmatrix} 11 \\ 29 \end{bmatrix}$, (e) [17]

22 (a) $\begin{bmatrix} 2 \\ 1 \\ 0 \end{bmatrix}$, (b) not possible, (c) $\begin{bmatrix} 1 & 0 & 3 \\ 2 & 1 & 1 \\ 1 & 0 & 1 \end{bmatrix}$,

(d) $\begin{bmatrix} 1 & 0 & 3 \\ 2 & 1 & 1 \\ 1 & 0 & 1 \end{bmatrix}$

23 $\begin{bmatrix} x + 2y \\ -x + 4y \end{bmatrix}$

24 $\begin{bmatrix} 0 & a^2 \\ b^2 & 0 \end{bmatrix}$

$$25 \quad \begin{bmatrix} 0 & 8 & 13 \\ 3 & -6 & -9 \\ 12 & 0 & 3 \end{bmatrix}$$

26 As given in the problem

$$27 \quad \mathbf{A}^T = \begin{bmatrix} 1 & 4 \\ 2 & 5 \\ 3 & 6 \end{bmatrix}, \mathbf{B}^T = \begin{bmatrix} 1 \\ 2 \\ 3 \end{bmatrix}, \mathbf{C}^T = \begin{bmatrix} 1 & 3 & 5 \\ 2 & 4 & 6 \end{bmatrix}$$

28 A sum of symmetric matrices is symmetric.

$$29 \quad \mathbf{A}^{-1} = \frac{1}{7}\begin{bmatrix} 3 & -4 \\ 1 & 1 \end{bmatrix}, \mathbf{B}^{-1} = \frac{1}{10}\begin{bmatrix} 2 & 2 \\ -1 & 4 \end{bmatrix}, \mathbf{C}^{-1} = \begin{bmatrix} 2 & -1 \\ -3 & 2 \end{bmatrix},$$

$$\mathbf{D}^{-1} = \begin{bmatrix} 2 & -5 \\ -1 & 3 \end{bmatrix}$$

$$30 \quad \mathbf{A}^{-1} = \frac{1}{40}\begin{bmatrix} 10 & 6 & 8 \\ 5 & -9 & -12 \\ 5 & 7 & -4 \end{bmatrix}, \mathbf{B}^{-1} = \frac{1}{2}\begin{bmatrix} 2 & 0 & -2 \\ -3 & 0 & 4 \\ 3 & 2 & -4 \end{bmatrix},$$

$$\mathbf{C}^{-1} = \frac{1}{2}\begin{bmatrix} 1 & -1 & 1 \\ -1 & 1 & 1 \\ 1 & 1 & -1 \end{bmatrix}$$

31 (a) $x = -3, y = 1$, (b) $x = -7/4, y = -1/4$,
   (c) $x = -13, y = -8$, (d) $x = 0, y = -1, z = 2$,
   (e) $x = -5/3, y = 5, z = -4$, (f) $x = 7/9, y = 13/9, z = 15/9$

Chapter 4

1 (a) -2, (b) -10, (c) 22, (d) -5

2 (a) 240, (b) -22, (c) -69, (d) 14, (e) 87

3 (a) $-8, -14, 4$, (b) $-12, -10, +6$, (c) $-8, -12, 17$

4 (a) -63, (b) 25, (c) 23, (d) 4

5 (a) 1, (b) 0, (c) 1, (d) $1 + a^4 + a^8$

6 Det $\mathbf{A} = $ det $\mathbf{B}$

7 (a) 1, (b) 1, (c) -1

8 2

9 -36

10 (a) 12, (b) 12, (c) 24, (d) -12

11 (a) 29, (b) 432, (c) 6

12 As given in the problem

13 As given in the problem

14 $-3, \pm\sqrt{3}$

15 (a) -6, (b) -13, (c) 6, (d) 1, (e) 3, (f) -5, (g) 14, (h) -2

16 (a) –27, (b) 4, (c) –1, (d) –78, (e) 4, (f) 3, (g) 56, (h) 6

17 (a) 144, (b) 20, (c) 40, (d) –232, (e) 160

18 2/3

19 As given in the problem

20 (a) –10, (b) 20, (c) 30, (d) –10, (e) 10, (f) 10, (g) 10

Chapter 5

1 (a) $\begin{bmatrix} -1 & 0 & 0 \\ 0 & 0 & -1 \\ 0 & -1 & 0 \end{bmatrix}, \begin{bmatrix} -1 & 0 & 0 \\ 0 & 0 & -1 \\ 0 & -1 & 0 \end{bmatrix}$,

(b) $\begin{bmatrix} 14 & -2 & -4 \\ -4 & 7 & 5 \\ 10 & 4 & 8 \end{bmatrix}, \begin{bmatrix} 14 & -4 & 10 \\ -2 & 7 & 4 \\ -4 & 5 & 8 \end{bmatrix}$

2 (a) $-\begin{bmatrix} -1 & 0 & 0 \\ 0 & 0 & -1 \\ 0 & -1 & 0 \end{bmatrix}$, (b) $\dfrac{1}{18}\begin{bmatrix} 14 & -4 & 10 \\ -2 & 7 & 4 \\ -4 & 5 & 8 \end{bmatrix}$

3 (a) $x = -1, y = 1, z = 2$, (b) $x = 2, y = -1, z = -1$,
   (c) $x = 3, y = 1, z = 0$, (d) $x = 1, y = 2, z = 3$,
   (e) $x = 2, y = 1, z = 1$

4 (a) $x = 1, y = 2$, (b) $x = 2, y = -2$, (c) $x = 3, y = -1$

5 As in problem 3

6 (a) $x_1 = 1, x_2 = -1, x_3 = 2, x_4 = 1$,
   (b) $x_1 = 1, x_2 = 2, x_3 = 1, x_4 = -2$

7 (a) $\begin{bmatrix} 6 & 0 & 0 \\ 0 & 3 & 0 \\ 0 & 0 & 2 \end{bmatrix}, \begin{bmatrix} 6 & 0 & 0 \\ 0 & 3 & 0 \\ 0 & 0 & 2 \end{bmatrix}$,

(b) $\begin{bmatrix} 1 & -2 & -3 \\ 0 & 1 & 0 \\ 0 & 0 & 0 \end{bmatrix}, \begin{bmatrix} 1 & 0 & 0 \\ -2 & 1 & 0 \\ -3 & 0 & 0 \end{bmatrix}$,

(c) $\begin{bmatrix} -4 & -2 & -10 \\ 1 & 2 & 7 \\ 2 & -2 & 2 \end{bmatrix}, \begin{bmatrix} -4 & 1 & 2 \\ -2 & 2 & -2 \\ 10 & -7 & -2 \end{bmatrix}$

8 (a) $\dfrac{1}{6}\begin{bmatrix} 6 & 0 & 0 \\ 0 & 3 & 0 \\ 0 & 0 & 2 \end{bmatrix}$, (b) $\begin{bmatrix} 1 & 0 & 0 \\ -2 & 1 & 0 \\ -3 & 0 & 0 \end{bmatrix}$,

(c) $-\begin{bmatrix} -4 & 1 & 2 \\ -2 & 2 & -2 \\ 10 & -7 & -2 \end{bmatrix}$

9 (a) $x = 4$, $y = 1$, $z = -2$, (b) $x = -1$, $y = 2$, $z = 2$,
   (c) $x = 2$, $y = 1$, $z = 2$, (d) $x = 1$, $y = 4$, $z = 3$

10 $i_1 = \dfrac{Z_2(e_1 - e_3) + Z_3(e_1 - e_2)}{Z_2Z_3 + Z_3Z_1 + Z_1Z_2}$,

   $i_2 = \dfrac{Z_1(e_2 - e_3) + Z_3(e_2 - e_1)}{Z_2Z_3 + Z_3Z_1 + Z_1Z_2}$,

   $i_3 = \dfrac{Z_1(e_3 - e_2) + Z_2(e_3 - e_1)}{Z_2Z_3 + Z_3Z_1 + Z_1Z_2}$

11 $u = 5$ m/s, $a = 3$ m/s^2

12 $5x + 3y = 11$

Chapter 6

1 (a) $V_1 = 4$ V, $V_2 = 1$ V, (b) $V_1 = 2$ V, $V_2 = -1$ V, $V_3 = 1$ V,
   (c) $V_1 = 5$ V, $V_2 = 2$ V, $V_3 = 1$ V,
   (d) $V_1 = 1$ V, $V_2 = -2$ V, $V_3 = 2$ V, $V_4 = 3$ V

2 (a) 1.14 A, 0.43 A, 0.71 A,
   (b) 1.82 A, 1.09 A, 0.91 A, 0.91 A, 2 A,
   (c) 4 A, 0 A, 2 A, 2 A, 2 A, 2 A, (d) 2 A, 3 A, 3 A, 2 A, 1 A

3 (a) $\begin{bmatrix} 0.2 & -0.1 \\ -0.1 & -0.3 \end{bmatrix} \begin{bmatrix} V_1 \\ V_2 \end{bmatrix} = \begin{bmatrix} 1 \\ 3 \end{bmatrix}$,

   (b) $\begin{bmatrix} 3 & -1 & -2 \\ -1 & 5 & 0 \\ -2 & 0 & 5 \end{bmatrix} \begin{bmatrix} V_1 \\ V_2 \\ V_3 \end{bmatrix} = \begin{bmatrix} -5 \\ 3 \\ -3 \end{bmatrix}$,

   (c) $\begin{bmatrix} 3 & -2 \\ -2 & 2 \end{bmatrix} \begin{bmatrix} V_1 \\ V_2 \end{bmatrix} = \begin{bmatrix} 1 \\ 3 \end{bmatrix}$

4 As the answers to problem 2

5 (a) $\begin{bmatrix} 9 & -3 & -4 \\ -3 & 9 & -1 \\ -4 & -1 & 8 \end{bmatrix} \begin{bmatrix} I_1 \\ I_2 \\ I_3 \end{bmatrix} = \begin{bmatrix} 0 \\ 0 \\ -6 \end{bmatrix}$,

   (b) $\begin{bmatrix} 3 & 0 & 0 & -2 \\ 0 & 5 & -3 & 0 \\ 0 & -3 & 7 & 0 \\ -2 & 0 & 0 & 5 \end{bmatrix} \begin{bmatrix} I_1 \\ I_2 \\ I_3 \\ I_4 \end{bmatrix} = \begin{bmatrix} 2 \\ 0 \\ -4 \\ 0 \end{bmatrix}$

6 (a) 0.2 A, 0.5 A, 0.7 A, (b) 1 A, 5 A, 4 A, (c) 4 A, 2 A, 2 A,
   (d) 0.35 A, 0.078 A, 0.40 A, 0.43 A, 0.32 A,
   (e) 0.5 A, 3 A, 2 A, 0.5 A, (f) 4 A, 1 A, 2 A, 3 A, 1 A, 2 A

7 (a) $\begin{bmatrix} 9 & -2 & 0 & -1 \\ -2 & 8 & -6 & 0 \\ 0 & -6 & 11 & 0 \\ -1 & 0 & 0 & 5 \end{bmatrix} \begin{bmatrix} V_1 \\ V_2 \\ V_3 \\ V_4 \end{bmatrix} = \begin{bmatrix} 0 \\ 1 \\ -2 \\ 2 \end{bmatrix}$,

(b) $\begin{bmatrix} 6 & -3 & -2 \\ -3 & 7 & -4 \\ -2 & -4 & 11 \end{bmatrix} \begin{bmatrix} V_1 \\ V_2 \\ V_3 \end{bmatrix} = \begin{bmatrix} 3 \\ -2 \\ 0 \end{bmatrix}$

8 (a) $\begin{bmatrix} 3 & -1 & 0 \\ -1 & 5 & -4 \\ 0 & -4 & 7 \end{bmatrix} \begin{bmatrix} I_1 \\ I_2 \\ I_3 \end{bmatrix} = \begin{bmatrix} 4 \\ 0 \\ 6 \end{bmatrix}$,

(b) $\begin{bmatrix} 5 & -2 & 0 & -3 \\ -2 & 7 & -4 & 0 \\ 0 & -4 & 15 & -6 \\ -3 & 0 & -6 & 9 \end{bmatrix} \begin{bmatrix} I_1 \\ I_2 \\ I_3 \\ I_4 \end{bmatrix} = \begin{bmatrix} 1 \\ -2 \\ 0 \\ -3 \end{bmatrix}$

Chapter 7

1 $\begin{bmatrix} 35 & -35 \\ -35 & 35 \end{bmatrix}$ MN/m

2 $\begin{bmatrix} 140 & -140 & 0 & 0 \\ -140 & 240 & -100 & 0 \\ 0 & -100 & 180 & -80 \\ 0 & 0 & -80 & 80 \end{bmatrix}$ MN/m

3 $F_1 = 7$ kN, $F_2 = -8$ kN, $F_3 = 7$ kN, $F_4 = 8$ kN

4 $\begin{bmatrix} k_1 + k_2 & -k_1 - k_2 \\ -k_1 - k_2 & k_1 + k_2 \end{bmatrix}$

5 $\begin{bmatrix} k_1 & -k_1 & 0 & 0 & 0 \\ -k_1 & k_1 + k_2 + k_3 & -k_2 & -k_4 & 0 \\ 0 & -k_2 & k_2 + k_3 & -k_3 & 0 \\ 0 & -k_4 & -k_3 & k_3 + k_4 + k_5 & -k_5 \\ 0 & 0 & 0 & -k_5 & k_5 \end{bmatrix}$

6 $u_{x2} = -0.20$ mm, $u_{y2} = -0.77$ mm, $F_{x1} = 20$ kN,
$F_{x3} = -20$ kN, $F_{y3} = 20$ kN. All other values 0.

7 (a) $\begin{bmatrix} k_1 & -k_1 & 0 \\ -k_1 & k_1 + k_2 & -k_2 \\ 0 & -k_2 & k_2 \end{bmatrix}$, (b) $\begin{bmatrix} k_1 & -k_1 & 0 & 0 \\ -k_1 & k_1 + k_2 & -k_2 & 0 \\ 0 & -k_2 & k_2 + k_3 & -k_3 \\ 0 & 0 & -k_3 & k_3 \end{bmatrix}$,

(c) $\begin{bmatrix} k & -k & 0 & 0 \\ -k & 2k & -k & 0 \\ 0 & -k & 2k & -k \\ 0 & 0 & -k & k \end{bmatrix}$, (d) $\begin{bmatrix} k & -k & 0 & 0 & 0 \\ -k & 2k & -k & 0 & 0 \\ 0 & -k & 2k & -k & 0 \\ 0 & 0 & -k & 2k & -k \\ 0 & 0 & 0 & -k & k \end{bmatrix}$

8 (a) $u_x = 0.45$ mm, $u_y = 2.12$ mm, (b) $u_x = -0.20$ mm, $u_y = -0.77$ mm, (c) $u_x = 0.12$ mm, $u_y = -0.09$ mm

Chapter 8

1 (a) $x = 2$, $y = -1$, (b) $x = 1$, $y = 3$, (c) $x = 1$, $y = 3$, $z = 1$, (c) $x = 1$, $y = 3$, $z = 1$, (d) $x = -2$, $y = 1$, $z = 1$, (e) $x = 3$, $y = 1$, $z = 2$, (f) $x = 1$, $y = 1$, $z = 2$

2 As given in the answers to problem 1

3 Diagonally dominant (b), (d), (e)

4 (a) $\begin{bmatrix} 4 & 1 \\ 1 & -2 \end{bmatrix}$, (c) $\begin{bmatrix} -4 & 1 & 2 \\ 2 & -5 & 1 \\ 2 & -1 & 4 \end{bmatrix}$

5 (a) $x = 2$, $y = 2$, $z = 1$, (b) $x = 1$, $y = -1$, $z = 2$, (c) $x = 1.5$, $y = 0.2$, $z = 0.8$, (d) $x = 1.2$, $y = 0.9$, $z = 1.5$, (e) $x = 0.2$, $y = 0.5$, $z = -0.1$

6 (a) 2.35 A, 0.17 A, 3.21 A, 2.52 A, 7.56 A, 3.04 A,
   (b) 1.2 A, 0.8 A, 0.2 A,
   (c) 0.31 A, 0.43 A, 0.41 A, 0.10 A, 0.08 A, 0.12 A,
   (d) 0.52 A, 0.49 A, 0.80 A, 0.03 A, 0.31 A, 0.28 A

Index